Y0-BRM-303

water
drops

excelsior editions
AN IMPRINT OF STATE UNIVERSITY OF NEW YORK PRESS

water drops

celebrating the wonder of water

Peter E. Black

Foreword by
Gerald E. Galloway

Published by
STATE UNIVERSITY OF NEW YORK PRESS
ALBANY

Printed in the United States of America

EXCELSIOR EDITIONS
is an imprint of State University of New York Press

For information, contact
State University of New York Press
www.sunypress.edu

Production and book design, Laurie Searl
Marketing, Kate McDonnell

Library of Congress Cataloging-in-Publication Data

Black, Peter E.
Water drops : celebrating the wonder of water / Peter E. Black.
p. cm. — (Excelsior editions)
Includes bibliographical references and index.
ISBN 978-1-4384-4487-1 (hardcover : alk. paper)
ISBN 978-1-4384-4486-4 (pbk. : alk. paper)
1. Water. 2. Hydrology. 3. Watersheds. 4. Water quality. 5. Water conservation. I. Title.

GB671.B55 2012
553.7—dc23 2012019606

10 9 8 7 6 5 4 3 2 1

I know of no safe depository of the ultimate powers of society but the people themselves; and if we think them not enlightened enough to exercise their control with a wholesome discretion, the remedy is not to take it from them, but to inform their discretion.

In keeping with these words of the third United States president, this work is dedicated to your and my descendants, who need to understand the science and culture of water to participate effectively in the important decisions about the policies affecting and the management of our water and related atmospheric, biospheric, energy, and land resources, to assure ecological, environmental, and human sustainability.

Contents

Preface

This collection of essays consists of the scripts (as recorded) for a three-year weekly series of two-minute (really ninety seconds) broadcasts by WRVO-FM, Oswego, New York, during breaks (on Saturday afternoons and repeated Sunday mornings) in NPR programming. I conceived of, proposed, wrote (and re-wrote!) these essays, some from scratch, some abstracted from more extensive material in my two books. David E. White, ESF Radio Coordinator helped by editing and making suggestions for the scripts, as did Christopher P. Baycura, Producer, who recorded and edited them with me in the recording studios of the State University of New York College of Environmental Science and Forestry in Syracuse, New York. WRVO Program Manager Fred Vigeant recorded the INTRO and TAG and he and his staff manage the connections with Public Radio Exchange (www.prx.org). As of this writing, the essays are broadcast from an unknown number of NPR stations, are on the Nebraska Department of Water Resources Web site, and similarly appear on Web sites of several NGOs.

Having spent my forty-plus professional years teaching (and conducting research, consulting, and writing) about water and related land resources, I retired in 2000 "from a job, not a career," and have continued a variety of activities (old and new, for me) in education about water resources. A particularly delightful, fascinating, and productive activity has been since 1985 representing SUNY ESF as an advisory member of the New York State Soil and Water Conservation Committee, where there were continual challenges to expound upon some technical issue relating to water. While much of the information I passed on was to farmers, I also gained a renewed respect for farmers: they are closest to and know the land best. As a consequence, I conjured up a healthy respect for having farmers on public watersheds where soils and other conditions are appropriate for farming. Farmers have an investment in the land and its resources, and thus have a vital concern and take a special pride in protecting the watershed values. They are the first to know about threats to the land temporarily entrusted to them, notably fire, windstorm, insect, weed, and invasive species outbreaks. Farmers have a vested interest that often is not present on public lands, even though we, the citizen "owners," ought to exhibit care about their management. Over the years, I have served on numerous other local, state, and national advisory groups, including water resource and watershed committees and task forces all the way to State Technical Committee of the Natural Resources Conservation Sevice and the Environmental Advisory Board of the U.S. Army Corps of Engineers. While such service gives one a true sense of the practical soil and water conservation challenges facing us, just being there is an educational experience itself, one I was delighted to bring back to the classroom.

The idea for the radio program occurred to me during the spring of 2005 while working on the third edition of *Watershed Hydrology*. I had submitted two short essays (on *Black Ice* and *Bad Weather*) that were published in the local paper after radio or newspaper reporters gave incomplete or erroneous definitions,often badmouthing weather, I thought of Public Radio as an ideal way to get those and similar messages across to the public. The essays are, in part, derived from portions my books *Watershed Hydrology* and *Conservation of Water and Related Land Resources,* and many are inspired by reference to complex—and often inadequetly and/or unidentified—topics on news and other radio or TV broadcasts. I was inspired by John Weeks, who offered weekly broadcasts, *The Nature of Things,* on WRVO that were mostly reminiscences as he visited the Rice creek Field Station (near Oswego) and Baltimore Woods (near Marcellus). I regularly listen to *Star Date,* which gave me a length estimate as well as the challenge of getting some often complex information into a really brief time slot, and responded to the challenge. I was inspired, too, by my father, Algernon D. Black, who for many years broadcast *Ethical Issues in the News,* fifteen minutes *live,* on WQXR in New York City. I enjoy writing, am an inveterate tweaker—constantly editing—and found a new venue for education. For me, public service—the third "arm" of the Land Grant College mandate—is perhaps the most rewarding of all, for it deals with practical problems and the application of cultural restraints on technically based watery challenges to the public's natural resource base.

With recordings of three of these ninety second essays, I auditioned in July 2005 at WRVO-FM hoping that they might be interested in the program. They were enthusiastically received. Broadcasts commenced on January 7, 2006 (script next page).

I wished to provide these programs—both audio and written—because I believe I have something to contribute to the management of our water resources, not for profit. The staff at WRVO agreed that they would like to air the programs regularly for a year, and renewed *Water Drops* for 2007 and 2008. They currently are broadcast on Saturday and repeated Sunday of the same weekend. Local environmental and/or engineering consulting firms have been generous sponsors as underwriters of the broadcasts on WRVO. I have truly enjoyed the challenge of writing and editing these essays, as well as talking with listeners about water challenges, and working with WRVO personnel.

Why did I stop at 168? I had aimed for three years of weekly programs, and threw in a few extras—as well as the two that are specifically directed with thanks for the WRVO support—to allow some choice. I suppose I could say too that I ran out of gas, er, water.

Numerical designation, draft, edits, and recording dates, and length, have all been removed from the working scripts. The scripts average 243 words and ninety seconds without music, INTRO, or TAG.

In sum, *Water Drops* is a joint project of WRVO-FM, and the State University of New York College of Environmental Science and Forestry in Syracuse, New York, and Peter E. Black,

Distinguished Teaching Professor of Water and Related Land Resources, Emeritus, at SUNY ESF. The purpose of these essays is to provide interesting and useful information on water to the public. With better understanding, citizens may participate in making more informed water resources management decisions. Each of the informative recorded essays is currently aired twice weekly. Broadcasting on WRVO-FM, a central New York NPR affiliate, is currently expected to continue for three years. While *Water Drops* is copyrighted (as a working document and book) the segments are/will be available from Public Radio Exchange (http://www.prx.org) for free access and use, with appropriate citation and credit. Complete information is available at PRX and at my Web site, http://www.watershedhydrology.com/.

Acknowledgments

I am particularly grateful to Christopher P. Baycura, Producer, who not only smoothed and edited the recordings (with me so I could see and ask for deletions or re-timing), but also provided wonderful feedback as we chatted about the recordings of the day. David E. White, Editor and Radio Coordinator and longtime talk radio program host, provided innovative wording and ideas for opening statements as well as tone and editorial assistance. President C. B. "Neil" Murphy—all three at SUNY ESF—provided continuing enthusiastic support of the project, especially finances for the professionals' services and time. I could not have done this project without their technical, moral, fiscal, and enthusiastic support. The entire staff at WRVO has also been supportive, and I extend especial thanks to Fred Vigeant, Program Manager, Kelly Olsen, John Kraus, Station Manager, and to Tom Herbert, Matt Seubert, and Bill Gowan for their support. WRVO's Mark Lavonier composed, performed, and recorded the musical button, and graciously contributed it to the project and to any and all users. I am also particularly grateful to many friends who have heard the segments and have offered me their suggestions, comments, encouragement, and constructive criticism and ideas. Finally, I am most thankful to Foreword author General Gerald E. Galloway and Front Cover Photographer Ray Mathews for their contributions to this booklet. Any errors of fact, pronunciation, or misrepresentations, etc., are of course mine.

Foreword

Water Drops in a Water World

The world's experience in the first decade of the twenty-first century has clearly emphasized the importance of water and the challenge the world faces in dealing with this precious resource. Developed nations have begun to recognize that water is a necessity in their quest to continue growth and maintain the living standards to which they have become accustomed. Poorer nations have long recognized that water and water-related infrastructure are fundamental to providing for human health and welfare and economic development. Too little water—droughts—and too much water—floods—continue to be problems faced by both groups. The uneven distribution of water across the earth's surface complicates the picture and the potential impacts of climate change add to the difficulties. The United Nations estimates that by 2015, at least 40 percent of the world's population, three billion people, will live where it will be extremely difficult to get enough water to satisfy basic needs. In recent years, major floods have taken the lives of hundreds of thousands and crippled economies of the places where they have occurred. Many believe that this demand for water may create conflicts among nations and the potential for water wars. Still others believe that jurisdictions facing water challenges will find that water issues may become a tool for greater cooperation among these jurisdictions.

The United States faces many challenges in dealing with its water. While most communities have built facilities that at one time provided quality drinking water to their residents, many of the facilities are reaching the end of their life span and are in need of major rehabilitation and upgrade to handle myriad emerging contaminants that we must now address. The American Society of Civil Engineers (ASCE) reports a greater than $11 billion annual backlog in required work on these treatment facilities. The quality of the water in our rivers and in our estuaries has yet to meet the lofty goals we put forth during the halcyon days of the environmental movement when we demanded that our waters be *fishable, swimmable, and drinkable.* Our waste water treatment plants face the same level of backlog in rehabilitation and upgrade and address only part of our water quality problems. Control of the massive input of contaminants into our rivers and estuaries from nonpoint sources of pollution such as city streets and farm fields continues to be marginally addressed. Alien invasive species continue to be brought into this country through a variety of carriers and we struggle to control their spread. Water shortages, once thought to be only a problem for the arid Southwest, can now be found across the nation as urban centers grow and require more water and the agriculture community recognizes the substantial increases in yields that can be obtained from irrigation. Conflicts among jurisdictions for access to available water has led to court cases, formation of water markets, and a significant reanalysis of how we are going to deal with water shortages in the long run. Over the last fifteen years, Alabama, Georgia, and Florida have sought without success to reach agreement on the allocation of waters they share and critical decisions are now being made by judges and not the states. Reductions in the flow of the Colorado River are forcing Californians to seek new sources of water far beyond their major cities. Groundwater is becoming an increasingly important part of our water supply, yet in many parts of the country we know little about its quality and quantity, and where it is located. Although since the early part of the twentieth century the nation has supported federal programs to reduce the impacts of floods, flood losses continue to grow each decade as the magnitude of floods increases and as people and property continue to be placed in or drawn to harm's way. Inland waterways in the

United States carry bulk cargoes such as farm products, coal, petroleum, and building supplies within the nation and to ports for export abroad, yet we continue to fall behind in the maintenance and upgrades of our waterways, ports, and harbors. While other nations invest heavily in ensuring that with the growth in world trade their facilities can handle the largest vessels and move the cargo within their nations in the most expeditious manner, we seem reluctant to take on this challenge. ASCE grades the current state of our dams, levees, wastewater and water treatment infrastructure, and our waterways at the "D" level.

We have also come to recognize that our efforts to build dams, channel rivers, levee floodplains, and clear land for agriculture and other uses has had a severe impact on aquatic ecosystems and our riverine and wetland environments. As a result, today we have underway multibillion dollar projects to restore the Everglades, the wetlands of coastal Louisiana, and the badly damaged ecosystems of the San Francisco Bay-Delta, the upper Mississippi and Missouri Rivers, and the Chesapeake Bay. We are also struggling to deal with a growing number of rare and endangered species. Clearly, we have our work cut out for us.

Many of our water problems result from a failure on the part of decision makers at all levels of government and the public at large to understand what water is, how much there is, how it is used, how serious our challenges are, and what we must do to provide for a sustainable future. Given our nation's wealth, we do little to deal with the significant water problems of developing nations and to reduce the continuous loss of life in those countries from the lack of access to water. It must be that we simply do not understand what is happening.

We operate in this country without a national water policy. We do not relate land-use to water use and the two are inseparable. We have not conducted an assessment of the status of our nation's water since 1976, and, as earlier noted, we have failed to provide funding for water-related infrastructure at the level required to sustain it over time. Much of the water infrastructure that we are using today results from the visionary work of early-twentieth-century officials who invested in water resource development to meet not only the immediate needs of their time but also to provide for the demands of those would come after them. Many of our environmental problems stem from a simple shortsightedness that suggested that nature could handle anything we might thrust upon it. Other environmental problems have resulted from our willingness to sacrifice the environment to support economic growth without recognizing that sooner or later we would have to go back and address the problems we were creating.

I can only imagine how much better the world and the United States might be today if our leaders better appreciated and understood water and the complexity of modern water resources management. How much better off would we be if a well-informed public had demanded of their elected officials attention to water issues and if they had concurrently assumed responsibility for their part in dealing with water?

Peter Black, in *Water Drops,* has provided us a much-needed guide to understanding water and water issues. He has put to paper his more than forty years of experience in the field of water management and given us his perceptive insights. If every other day a family that is sitting down to dinner would read one of his essays and spend a few minutes at the table discussing its implications in their lives, over the course of a year we would produce individuals and family units that recognize and are motivated to do something about our water challenges. Politicians, businessmen and businesswomen, educators, doctors, government officials, and the public at large would be better able to carry out their work because they would understand how their actions would affect or would be affected in the long and short term by our approach to water. It is not too late. Those who read *Water Drops* will be developing this better knowledge of water and will help to make this twenty-first century far better than the twentieth in the way we address these important water challenges. I hope they will share their *Water Drops* experience with others and help to build an informed America. I only wish we had started sooner.

Gerald E. Galloway, PE, PhD
Washington, D.C., 2009

INTRODUCTION

Water has fascinated me for a long time. I have spent my whole adult life studying and teaching about water in our environment. I will share with you my wonderment of water and some fun and unusual things that you can see for yourself. Water is really a most astonishing substance. It exists at normal temperatures as a gas, liquid, or solid. It is the basis for life on this planet, and almost all of it is in the oceans. That's the reason Earth looks like a blue marble in the wonderful pictures from space. We look for water on Mars and other planets because we believe that is how we'll find life elsewhere in the universe. Looking inward, the human body is about 70 percent water and our blood is about 95 percent water. Water cools us in hot weather, removes waste products, and helps move nutrients around. Since water gains and loses heat energy slowly, it also helps regulate our body's temperature. If we perspire too much we suffer dehydration. In fact, we can live a lot longer without food than we can without water. We also need water to grow food; manufacture products; transport goods around the world; produce electricity; fight fires; clean our neighborhoods; grow lawns and gardens, and to have fun too. We all spend our first nine months in water. Sometimes, we just like to sit and look at water. The "water drops" in this book are presented to celebrate the wonder of water.

Science

Water on Earth

Do you think our planet should be called *Water* instead of *Earth*? Our ancestors walked Earth's solid surface, and thus called it "Earth." Early space travelers noted that Earth appeared blue because about 70 percent of Earth is covered by ocean. Actually, almost the entire planet is covered with water. No, I'm not crazy: the poles (and many mountains) are covered with snow and ice, and the atmosphere contains lots of clouds and water vapor. Ninety-seven percent of Earth's water volume is salty. Only 3 percent is fresh; about 2 of those 3 percent are icecaps (which change as our climate changes). Three-quarters of the remaining 1 percent is groundwater, and nearly all the rest is in lakes. The remaining fraction of a percent circulates between air, plants, animals, soil, and streams and rivers. So, almost all the water on the planet is in storage; only a small percentage of it is fresh, and only a tiny percentage of that is accessible for our use. But, all the water is important to us in many ways, quenching our thirst, helping plants grow, absorbing energy, gases, and lots of different substances, such as dissolved rock, animal and plant wastes, and space debris: the waters of the Earth both give us life and protect us. We may have to revise our thinking about how we manage our precious water so that we can continue to have the water we need available whenever we need it.

Ocean

We each need to drink four to five pints of water every day to survive, but we need all the other water on the planet, too, especially the oceans.* Our oceans, greater in depth than Earth's tallest mountains, and covering two-thirds of the planet's surface, absorb energy, gases, products of erosion and corrosion of Earth's solid matter, meteorites, and waste materials from plants and animals. They provide a medium on which ships can float, and they help equalize the heat energy around the planet. Ocean waves make beaches and erode coastlines. In addition, oceans provide abundant plant and animal life that give us food, medicines, a great diversity of ecosystems, sporting events, and beautiful scenes. Scientists affirm that there are not likely to be extraterrestrial forms of life without water on a planet that has a near-circular orbit; that is, not too far from a sun that is about the same age as ours, in a star system that has a big planet nearby (like Jupiter) that can attract asteroids and other space debris so as to protect our planetary spaceship. Also, all chemical reactions—including natural and man-made nuclear explosions—take place in the presence of water. We don't just need those few pints of water each day to survive; we need them to thrive. Aquatic environments are the foundation of life as we know it; we would not survive unless the conditions just described exist.

---

*We often refer to the total salty portion of the hydrologic realm as "oceans," but there is, in fact, only one ocean on the planet, and while it has multiple names—Atlantic, Pacific, Indian, Arctic—that doesn't segregate the water involved!

Biodiversity

Biodiversity is the sum of all animal and vegetable inhabitants of Earth's biomes, regions defined by similar climate, animals, and vegetation, and their relationships with one another. Biomes include tundra, deserts, grasslands, forests, wetlands and estuaries, and oceanic environments. There are more than thirty different kinds of biomes, including rainforests and coral reefs. Few are obviously water based, but all are dependent on or defined by H_2O. Humans, the most invasive species, survive to some extent in all of these environments, the oceans excepted. If we alter biomes, these complex systems on which we depend directly for food, clothing, or shelter, and for habitable conditions, all life changes. It is not just a matter of preserving members of every species, but of ensuring the variety of forms making up our life-support systems. We depend on the pliability of Earth's diversified biomes, the variety of life that provides protection against disease, plague, fire, and flood. The collective interactions of all the parts ensure our survival: if environments are altered, if there are changes in temperature or in distribution of ice, water, oxygen, or carbon, then the conditions under which life is sustained change. Water plays such a central role—in the energy cycle; in gaseous, waste, and nutrient exchange between plants and animals; in climate and weather; and in all aspects of our lives from conception to death and decay—that biodiversity is not only an appropriate subject for a Water Drop essay, I would be delinquent were I not to include it. Biodiversity is the most important wonder of water.

Gaia

Water is a major role player in the concept of Earth as a living system, a concept called *Gaia*. Water has positive and negative feedback mechanisms that cool Earth off if it gets too hot, and vice versa. Negative feedback mechanisms restore the condition that changed. Indeed, water is the all-important substance for life's existence. Without it, we would not be here. The Gaia concept is based on thinking of Earth's ecology as a living system, not an organism (although Earth behaves as if it were one). The concept is that species and their environments develop or evolve together. If one changes dramatically, so will others. The concept and details of Gaia help understand how Earth's life systems work. John Lovelock—who wrote *The Ages of Gaia*—was on a team of scientists engaged in the search for life in the universe by determining where and what to seek on other planets. In order to do that, it was necessary to consider how life developed on our own planet. I used the word *develop* because I know that not everyone agrees about life being by accident or design. The crucial feature of the Gaia concept is that species and their environments affect one another. However life arrived here, we are part of Earth's system now. It is all we have, and if we can understand it better, hopefully we can do a better job of managing it for our sustainability.

Tipping Point

Many of Earth's biodiversity systems are in danger, and are already changing. All the changes are reflected in Earth's water resources. These water-based systems include polar ice and mountain glaciers, and ocean coral and life forms. Symptoms include ocean acidification; melting of tundra; fires; and the wild and changing weather. All are symptoms of the global warming that we have caused, primarily by burning fossil fuels during the Industrial Revolution. These changes in biodiversity are not good for us. Humans really do not know how much of Earth's essential life support systems we need for sustainability. We do know, however, that many of Earth's ecological niches are changing dramatically. We do not know if there is a "tipping point," a level of some support system beyond which the systems may not recover. If a tipping point—a point of no return—does not exist, then humankind need not be concerned over immediate counteracting measures. However, if such a tipping point does exist, we need to pay immediate attention. That may be a Good Idea in either case. All citizens need to participate in deciding how we are going to maintain Earth's watery life support systems. As the universal and fundamental environmental substance, water is basic to all Earth's biodiversity. The almost daily changes are alarming to many scientists, as well as victims of storm, fire, flood, and drought, major causes of life-threatening conditions and personal loss, all water related.

CO_2, O_2, and H_2O

Earth's life depends on water. Carbon dioxide, oxygen, and water vapor make life cycles possible. Water is, of course, the principal agent that moves carbon dioxide and oxygen around, and is a gas itself in the vapor state. Plant leaves release oxygen *to* and take carbon dioxide *from* the atmosphere at the water surface in the tiny openings called *stomates* or *stomata,* which are usually underneath the leaves. The relative humidity of the air—that is, the percentage of water vapor that the air could hold at the current temperature and pressure—controls how much the stomate's guard cells open to allow the exchange of the two vital gases. Opposite the plants, animals *breathe in* the oxygen and *exhale* the carbon dioxide. *That* exchange takes place in the *lungs*, also in the presence of water in the cells that make up the tiny lung sacs. Blood is mostly water, absorbing oxygen from the lungs for your brain and other organs, and returning carbon dioxide to the lungs to be breathed out. The energy to move the water and its gases around comes from the sun, which also evaporates the water that cools us as well as helping move the vital life gases between plants and animals. Of course, cooling evaporation also occurs in the oceans, streams, rivers, ponds, lakes, and wetlands. And plants use carbon dioxide to build carbo*hydrates* (water carbon molecules) that build the cells of plants and the bones of animals. How wonderful!

Unusual

Water is truly an amazing substance. For example, the word *water* enabled deaf mute Helen Keller to understand what language was and therefore to speak: it opened an entire world for her. Water also has very unusual physical properties: it exists naturally on Earth in all three states: liquid, solid, and gas. Water is least dense at just above freezing, enabling aquatic life forms to survive killing winter temperatures under the floating ice. The vapor pressure over ice is less than it is over water just above zero degrees Centigrade, so that ice crystals attract water vapor. That enables the condensation and precipitation processes, soil porosity, and snowpack crusts. The water molecule is polarized and thus can combine with a great variety of compounds, making it a universal solvent except for the noble metals. The water molecule may be tracked with extra hydrogen as *heavy water* or *deuterium.* And water can also contain a stable isotope of the oxygen atom, useful in tracking water movement and storage in the hydrological cycle, the basis for isotope hydrology. So different proportions of isotope and regular water will aid in determining how long the runoff has been in storage. Finally, all chemical reactions take place in the presence of water, vital to life processes and industry. All in all, water has some very unique properties, and we would not be here without it or some conceivable substitute.

Circulation

Water circulates in Earth's environment much as it does in our bodies in the form of blood. Blood is about 90 percent water. And that global circulation provides some of the same functions on which our individual lives depend. Forming and falling through the atmosphere, water drops—and to a lesser extent snow crystals and hail—absorb gases present in the atmosphere, especially carbon dioxide, oxygen, and nitrogen. Water is a nearly universal solvent—its molecules can connect with almost all others except the noble metals—so water is definitely *not* as pure as rain. Carbon dioxide makes water slightly acid; thus, the rain and runoff water can corrode rocks, soil, and painted surfaces. The substances dissolved in runoff provide nutrients for plants and animals that live in the water, especially oxygen for the animals, carbon dioxide for the plants, and silica from rocks and soil for the cell walls of plants. Nitrates, which form whenever lightning occurs, are nature's natural fertilizer. Evaporation from water bodies and transpiration from plants provide the return portion of the circulation system, and are similar to our blood circulation system. Both processes require energy. Energy of water at higher elevations is dissipated as it flows downhill to wetlands and the oceans, which functions as the Earth's recycle bin. Elimination from our bodies includes dissolved substances, but the evaporated ocean water, of course, contains none. What a wonderful system!

Water and Energy

Water is important in natural and artificial energy systems. The most important natural energy characteristic of water is its very high specific heat, the highest of any common substance. Specific heat is the amount of energy necessary to raise the temperature of something one degree Centrigrade. That makes water a great place to store heat energy. The oceans store lots of heat energy, moving it around. Water exchanges energy with air, resulting in fog if the air is colder than the water and evaporation if the air is warmer and drier. The oceans curb Earth's climate extremes with their tremendous amount of water: hurricanes or typhoons move heat energy poleward, especially in summer and fall. Another relationship between energy and water is expressed in the generation of electricity by building a dam and reservoir and routing water through turbines turned by the falling water, thereby converting that energy to electricity. Humans used falling water even before electricity became common by building waterwheels that helped grind grain and drove early factory manufacturing or processing operations. Water is also of considerable importance to nuclear power plants in three ways: First, heavy water (water enriched by the deuterium isotope) is necessary in the nuclear reaction. Second, water is used to convert the heat energy of the controlled nuclear reaction to electricity by driving turbines. And third, an immense amount of water is needed to cool the power plant. Water also plays major roles in the production and use of fossil fuels and geothermal power sources.

Water Vapor

Whenever the outdoor temperature is really cold, you can conduct an experiment when you shower. It will let you see the effect of temperature on water in the atmosphere. First, get a glass filled with ice, and then fill it the rest of the way with water, making sure that the glass is dry on the outside. Then, set it safely on the sink or floor. When you are through with your shower you may see a cloud of water vapor in the bathroom. If you don't and want to see one develop in a hurry, open the window, but dry yourself first so you don't freeze! A cloud develops as the frigid air pours through the open window and into the moist, humid bathroom air. Now look at the glass. It should have water droplets on it, and may be completely covered with water. When the relatively dry cold air poured though the window, it chilled the air in the bathroom and the moisture in the air condensed into tiny droplets to form the cloud. The water droplets formed on the glass because the gaseous moisture in the bathroom air condensed on its cold surface. In the atmosphere, the air is cooled by lifting it to a cooler level. Once drops start to form, they collect into larger and larger drops and become heavy enough to fall. Sometimes, they never get heavy enough and evaporate—become vapor again—before reaching the ground. That type of rain is called *virga*.

Evapotranspiration

When it comes to water, what comes *down*, goes *up*. In the water cycle, the words are *precipitation* (rain, ice, or snow), and *evaporation* (literally, making liquid water vaporize). That takes energy and, as you learned in an early science course, evaporation is a cooling process. That's why, along with the shading from the sun, the air in a forest is cooler. If the forest is cut, the air is hotter because the shade is gone and because *transpiration*—that part of the water cycle where trees move soil water through the stem, branches, and leaves—is gone. Transpiration occurs when water evaporates from the tiny openings called stomates or stomata in the leaf undersides. A full-grown tree can transpire lots of water from the soil: thus, houseplants keep the inside air more comfortable by being moist as well as cool. Nutrients in the water help the tree grow. And, since plants give off oxygen, inside air is both healthier and cooler with plants around. On the average, *evapotranspiration*—that is, evaporation and transpiration combined—lifts about twenty-two of the thirty inches of yearly precipitation back to the atmosphere—about 70 percent of the annual rain and snowfall. The rest runs off in streams and rivers. So, having park and forest vegetation around is quite important to your health and the natural functioning of Earth's water cycle. Even your cut Christmas tree continues transpiring water, which is why the tree stand has to be filled often.

Acid Rain

Recently there has been a lot of talk about "acid rain." It certainly sounds like it is a bad thing. But, actually, mildly acid rain is natural. It dissolves minerals and rock and keeps the Earth recycling itself. Four of the most abundant gases in Earth's atmosphere are *nitrogen*, *oxygen*, *water vapor*, and *carbon dioxide*. When water vapor molecules combine to form liquid drops, the least abundant of those four gases—carbon dioxide—dissolves in it readily. When that happens, it combines with some of the hydrogen atoms to form *carbonic acid*. Carbonic acid is pretty weak; we use it to help wash out our eyes. So, natural rainfall is slightly acid and has a *pH* or *reaction* of about 5.7 instead of 7.0, which is neutral, halfway between acid and base. Nearby or upwind smokestacks may release sulfur coal waste that puts sulfur dioxide in the atmosphere, too, and that makes rain with lots of sulfuric acid. Sulfur dioxide can make the pH of rainwater much lower, as low as 3 or 4. Acid that strong can move—or mobilize—aluminum from soils, and that can poison streams or lakes. That's real acid rain! You can buy an inexpensive pH test kit at your local aquarium supply or science store. It has paper strips that you can use to check the pH of the rain, snowmelt, your backyard pond, or even a puddle. It will also explain the pH scale. Try it!

Studies of Water

I called my own area of interest in water *water and related land resources*. This covers the broad areas of watershed hydrology, watershed management, and soil and water conservation policy. My underlying scientific interest was—and is—in the underlying ecology of water. And my cultural interest was—and is—in how we manage our need and desire for water in our everyday lives, including the law, policy, organizations, economics, programs, and strategies for making water available to all. There are a large number of specific areas of study of water. That is fitting, of course, since water is so prevalent on this planet and so much a part of—and actually underlies—life. In alphabetical order they include: aquatic ecology, aquatic entomology (about bugs in water), climatology, cryology (the study of ice in the solid form of ice), engineering hydrology, estuarine ecology, fisheries, geology, glaciology, groundwater hydrology, hydraulics (the behavior of water in pipes and channels), hydrology (the study of water in natural and disturbed environments), isotope hydrology (using radioactive chemicals to trace and time water movement), limnology (the study of fresh water), meteorology, oceanography, chemistry of water, physics of water, wetlands ecology, and wetlands hydrology. And there are interdisciplinary areas, such as water transportation, water in industry, water and energy, and water in art, planning, health, law, and literature. Watery careers are open to your imagination, creativity, and passion.

Measuring Water

When you need water for a bread recipe you use a tablespoon or a cup. But to measure water in the environment we need larger units. An inch of rain on your lawn would produce one inch of water in a straight-sided cup used to measure the rain. We measure the *depth* of rainfall in inches or millimeters, ignoring the total *volume* of water: the total water volume, of course, depends on whether the inch falls on your backyard or on a town one mile square. An inch of rain on a square mile would be over four billion cubic inches, too big a number to deal with. So, we use the acre to measure area. An acre is more than forty thousand square feet. Rain an inch deep on an acre would fill more than 65,000 five-gallon pails. If that were gasoline, it would fill your car's twenty-gallon gas tank almost three hundred times. On your quarter-acre backyard, an inch of rain would produce more than ten thousand cubic feet of water and you would need more than 16,000 five-gallon pails to hold it all! Figure *that* in cups or tablespoons! An inch deep in a cup or raingage is pretty small, but as rainfall on your lot, it can be a lot more than you can drink!

More Water Measurements

You can measure rainfall in your backyard. Inexpensive rain gauges are available at science, garden, and hardware stores. Locate yours four or more feet above the ground, with no obstructions in an imaginary inverted cone about forty-five degrees above the gauge. Trees and roofs can artificially affect the rainfall. The gauge should be kept clean and empty between storms. The amount of rainfall is measured by its *depth*—for example, in inches. A ski area's snowpack is measured by its total depth, but a hydrologist is interested in how much water is in the pack, reported as depth expressed as a percentage or in decimal figures as *density*. For example, a five-foot-deep snowpack with 50 percent water content—a density of 0.5—contains two and one-half feet of water. In the soil, the amount of water is reported as the depth in number of inches or as a percentage of the volume. Soil moisture is a lot more complex to measure than rainfall or streamflow. In a stream, the *discharge* is reported per unit time, as in cubic feet or cubic meters per second. A streamflow of one cubic foot per second yields 86,400 cubic feet per day, or one times the number of seconds in a day. At seven and one-half gallons per cubic foot, that is 648,000 gallons, enough for a town of more than four thousand people using 150 gallons per day per person.

Hydraulics

Hydraulics is the science of fluid mechanics in confined spaces, such as pipes, flumes, canals, streams, and rivers. It combines science and engineering, embraces fluid mechanics, and has been of alluring interest for thousands of years. Contrary to what we learned in school, the pressure of water everywhere in a closed container is *not* always the same. Even when moving, water pressure varies, as does the velocity. And water's mass means that it will be under greater pressure lower in the closed system. Many famous people have contributed to understanding water conveyance and behavior situations. An intriguing example involves sediment suspended in moving water. The size of the particle that moving water can suspend depends upon how fast the water is moving, and the speed of the water depends upon the slope of the channel. If one adds more and more sediment particles to water flowing in a channel, the water's velocity naturally limits how much it can carry: some sediment particles fall out, building ridges of ripples on the floor of the channel. The water can now flow faster on the downstream side of the ripple, thus can carry more sediment particles. As the water picks up more particles the slope decreases, and so does the velocity, repeating the cycle over and over, making the sediment ripples move downstream. That's just one of many interesting experiments in hydraulics. It is really a fascinating area of study and professional activity.

Wonder

Watching snowflakes on a late fall morning against a backdrop of dark trees from which the leaves had fallen, several ecological connections between trees and water occurred to me. I thought how neat it was that snowflakes float softly to Earth's leaf-covered soil surface during the winter, whereas spring and summer raindrops are heavier and fall faster, sometimes quite violently. I thought about how water plays important roles between the precipitation processes and tree growth and structure. A layer of cells at the base of a deciduous tree's leaves weakens as fall approaches, broken by wind or possibly when cell water expands to ice as temperatures drop to freezing. Also, deciduous trees' branches are protected from the change in precipitation from rain to snow. The snow, of course, collects on branches, but the absence of leaves dramatically reduces surfaces on which snow can rest. Also, most evergreen and deciduous trees have different woody cell structure, and the horizontal limbs of evergreens can withstand the heavier load of accumulated precipitation in the form of snow more readily than the often upward-angled deciduous tree branches. If early, heavy, wet fall snow falls on deciduous trees when leaves are still in place, it can lead to downed limbs and sometimes downed power lines. Of course, the fallen leaves decompose, recycling nutrients through soil and trees—with the aid of water as rain and snowmelt. The accumulated snow melts more gradually, seeping slowly into the soil for growth in the spring.

Water Storage

Unless you are hot or have been exercising, you normally drink only a small percentage of the water in your glass. Nature does that too. In fact, nearly all the water on the planet is in storage and may not move for a long, long time. As mentioned previously, 97 percent of Earth's water is stored in the oceans, circulating in the vast basins, and returning to the atmosphere by evaporation. Only 3 percent is fresh, and two-thirds of that is stored in ice, or was before recent warming. That leaves only 1 percent, three-fourths of which is stored in groundwater. Twenty percent of what's left is stored in lakes, leaving a tiny percentage circulating for our use. In storage, water dissolves storage medium materials and thus can be identified by its quality, allowing us to identify where the water came from and how we might use it. For example, groundwater is, typically, fifty-five degrees Fahrenheit, and thus is valuable for cooling. Most water in groundwater storage has passed through overlying soil, so when we build subdivisions with lots of pavement, roofs, and compacted lawns, we bypass that storage, flushing rain and snowmelt rapidly to nearby streams. The results are often severe: natural vegetation changes with less water available for plant growth; temperatures increase because natural evaporation cools the air; and the land can no longer support itself, so there may be severe sinking or subsidence.

Hydropedology

Hydropedology is a new field of research that studies the critical zone where soil and water interact at the surface of the Earth. Think about how important that critical zone is, and what it is like. For example, think about why the relentless energy of ocean waves continually hits beach sand without great change from day to day. That massive amount of energy—a force against which we cannot stand (we cannot even stand against a waist-high flowing current of more than about four feet per second)—is dissipated by the billions and billions of grains of sand. Each responds to the wave-crashing energy individually and interactively with other sand grains, perhaps just as the incredible numbers of stars—and comets, asteroids, and dark matter—absorb and maintain the energy of the universe. That brings to mind a paradox: in sand and soil there is overall stability based on continual change. The same concept applies on dry land, where there are similarly billions and billions of soil particles, plus organic material, that make up our familiar soil. Here the critical zone is the essence of soil and water conservation. Wind and rain and the mind-boggling power of water expanding as it freezes maintain the myriads of channels for the movement of water that would otherwise erode the soil as it moves from atmosphere to terrasphere.

Fertilizer in the Rain

Driving along a four-lane highway you might see stripes of dark green grass on the opposite lane's shoulder. That is water chemistry in action. Water is a universal solvent: almost everything can dissolve in it. During a thunderstorm, lightning combines nitrogen and oxygen molecules—the two most common gases in the atmosphere—to form nitrogen oxide, commonly called *nitrate*, which quickly dissolves in raindrops. We use man-made nitrate as a plant fertilizer to make our crops, lawns, and gardens grow better, but in the natural environment, nitrate can be provided by rain. So, during a summer thunderstorm, rain with lots of nitrate in it collects on highway surfaces and runs to the lowest spot, usually an expansion crack in the pavement. Then it flows down and over the highway shoulder, fertilizing the grass that grows there. So, next time you travel along a four-lane divided highway in mid to late summer, glance at the opposite lane's shoulder, and see if you can't see the regularly spaced stripes of dark green. If you do, you are actually seeing some water chemistry that is naturally important in our environment at work. You may be able to see the effect in other places, too, such as where concentrated rainwater regularly runs off other solid surfaces such as patios, driveways, or sidewalks, and fertilizes the nearby grass. Gravel under the pavement or sidewalk would allow the water to seep downward, avoiding this problem.

Plants and Water

All living plants require water, but some plants have special relationships with their environment; you probably know some of them. Mosses inhabit moist environments, especially wet sites, woodlands, and water bodies; epiphytes get their water directly from the air in which they live, not from soil; pitcher plants, the Venus Flytrap, and other bog plants form containers for water that trap insects and extract nutrients from them; the palo verde tree loses its leaves during a drought but its green bark can carry on photosynthesis without the typical transpiration loss of water through the leaves; coast redwoods thrive in very foggy environments and get much of their moisture from water vapor that readily condenses on their tiny diameter needles; the saguaro cactus, which can grow as tall as twelve meters—thirty-six feet—can live to two hundred years in the desert where its stored water enables almost yearly flowering. In humid regions, the invasive water hyacinth floats because of an air-filled honeycombed bulb. It can completely cover nutrient-laden waters. It is utilized to take toxic materials out of graywater (wastewater generated from domestic activities such as laundry, dishwashing, and bathing) in water-treatment plants, even as its invasive nature spreads it widely, and its decay depletes the water's oxygen content. In general, of course, plants take in carbon dioxide from the atmosphere and release oxygen, exactly the opposite of what animals do. Thus, we could not survive without plants that enable us to breathe life-giving oxygen.

Invasive Species

Purple Loosestrife, salt cedar, alewives, lamprey eels, Eurasian milfoil, and zebra mussels are examples of invasive species living in, near, or moved by water. But Earth's most invasive species is *Homo sapiens*. We fit all the criteria, including the ability to rapidly reproduce, live in and modify different environments, and compete successfully for living space. As long as we can move around and have fresh water and oxygen with us, we can survive anywhere, even on the moon. We are also exploitive and aggressive. Actually, many so-called invasive plant and animals species are "unintended consequences." Some plants were intentionally imported as exotics to show in parks, botanical gardens, or front yards. Others occur without our help, but the majority of them spread because we either intentionally or accidentally transport them—in ship ballast, attached to boat hulls, in or on our food, clothes, and commerce—as our products and vessels travel around the world. We delude ourselves with the term *invasive,* depicting us as victims when we are, in fact, the principal perpetrators. It would be a good idea to identify the factors that make species invasive and to focus on our role in the process. Put another way, it is irresponsible and probably quite unproductive to consider how to control invasive species while ignoring our role in their invasive status: we should focus on how we might control the factors that make plants and other animals invasive. *They* are innocent. *We* are not.

Mosses and Water (R. W. Kimmerer)

Mosses abound, and are now better known through my colleague Professor Robin Wall Kimmerer's book *Gathering Moss*. With her permission I have abstracted her description of the wonder of water at the centimeter-high moss scale. "The atmosphere is possessive of its water. While the clouds are generous with their rain, the sky always calls it back again with the inexorable pull of evaporation. . . . Every element of a moss is designed for its affinity for water. From the shape of the moss clump to the spacing of leaves along a branch, down to the microscopic surface of the smallest leaf; all have been shaped by the evolutionary imperative to hold water. . . . Water has a strong attraction for the small spaces in a clump of moss. The water molecules attach themselves readily to leaf surfaces, due to the adhesive properties of water. One end of the water molecule is positively charged, the other is negative. This allows water to adhere to any charged surface, positive or negative. . . . The bipolar nature of water also makes it cohesive, sticking to itself. As a result . . . water can form a transparent bridge between two plant surfaces. . . . This mossy sponge drips a constant flow of water to the tree roots, saturating the ground below and filling the soil reservoir." That's a poetic description of how water supports life. The ecological affinity of mosses for water conserves it while creating a favorable environment for these small plants.

Water and Trees

Water has some very complex interactions with trees. Some are straightforward, as when water supplies nutrients for tree growth. Trees protect the soil by intercepting raindrops and hail, reducing their impact and thus maintaining soil pores so that the water runs off slowly. Trees pump water from the soil to the atmosphere through *transpiration,* that is, evaporation through tiny openings on the undersides of leaves called *stomates* or *stomata.* That cools the air, as does the trees' shade. Interception by tree limbs and foliage also allows time for water to evaporate from them, thus keeping water from reaching the soil. Of course, if all the rainstorms were short and small, then so much would be intercepted that none would even reach the soil. But then we wouldn't have enough water to grow trees in the first place. Some trees also lift water from lower soil layers by transpiring during daylight hours. At night, the water settles from the trunk to upper soil layers, thus keeping surface soils moist for the next day. One great physics puzzle is how trees can lift water as high as three hundred feet in the coast redwoods, when atmospheric pressure can only lift water about thirty inches. There have been many theories. Finally, trees protect against floods from forested lands because they transpire so much water that there is plenty of storage space in the forest soil, even after a storm.

Interception and Energy

The word *interception* may bring to mind a football for you, but in the world of water interception occurs when rain or snowfall hits vegetation before reaching the soil. It happens on artificial surfaces too, such as buildings. In a forest, intercepted precipitation may be by tree crowns, understory vegetation, and vegetative forest-floor litter. Some intercepted water runs off saturated surfaces to the soil, but most evaporates from plant surfaces. That takes energy, from warm winds, or by solar radiation after the rain or snowfall. The complex characteristics of energy and the evaporating surfaces were written about in a classic 1965 Forest Service publication, *Radiant Energy in Relation for Forests,* by Reifsnyder and Lull. Important observations from that report include understanding that different surfaces in a forest provide opportunity to intercept and store snow and water; that the energy available for evaporation and transpiration is quite difficult to measure; and that changes in temperature—and therefore energy—occur during the intercepting and evaporating processes. The angle and aspect of the slope where the forest grows, its latitude, and the season of the year all affect available solar radiation. In California's coast redwood forests, daily ocean fog wets the massive amounts of foliage almost every night, so the next morning's solar energy evaporates the intercepted moisture, conserving water in the soil. Thus, the trees grow for a long time, and are quite tall. And, just to complicate the issue, remember that evaporation is a cooling process.

Interception Amount

Vegetation protects the soil from rain drop erosion. *Interception* is the term scientists use to describe the process in which the falling motion of rain and snow is interrupted and redistributed by vegetation. Interception thus reduces water going to soil moisture and runoff. Evergreens—mostly conifers—intercept precipitation year-round and, if the branches slope outward like those of the Norway spruce, very little water reaches the ground directly under the tree. Deciduous trees hold very little snow in winter, when the leaves are gone, but in summer they may temporarily store up to a quarter of an inch of rainfall on the leaves until it evaporates. Four days of quarter-inch rain storms could produce no water reaching the soil at all. Woodlot and forest stands often have an understory of brush or younger trees, which also intercept precipitation. So does the vegetative litter on the forest floor. Often an inch of rainfall is necessary before wetting the soil under hardwoods in summer. Some species of trees and brush have upward-sloping branches, and these funnel the water to the trunk where it runs very efficiently as stemflow to soil and roots. Coast redwoods (*Sequoia sempervirens*), have such branches, but the foot-thick bark of a mature tree is so porous that it can store all the stemflow for an entire year. If you lean up against the bark of one of these trees, you'll get quite wet.

Soil Storage

I've heard people call the forest floor "a big sponge" because it soaks up rainfall and snowmelt, but soil in the forest, pasture, or your backyard garden is not a sponge, although it behaves like one. Here's why. A sponge is a natural or artificial material with connected or unconnected holes or pores, whereas a soil is air filled with particles of sand, silt, and clay. Usually, the pores are connected, and water flows readily through the soil. If larger pores are filled with smaller particles, then water doesn't flow easily. For example, water doesn't move well in clay soils because both particles and pores are very small. But water does flow readily through larger pores in sandy soils. Between clays and sands are loams: mixtures of sand, silt, and clay. Water is stored in micropores—also called *capillary pores*—for long periods, and in macro- or noncapillary pores for short periods. Tiny capillary pores hold water so tightly that water can't be pulled out of the pores by gravity, which is always present. Water in the noncapillary pores can flow to groundwater or to streams. Plants and animals in soils are important: they keep the soil porous so that water may move into and through it. Ice is also important in soils. As water freezes, it expands. So, when soil moisture freezes, it makes the pore spaces larger, providing for more soil storage.

Frost in Soil

Frost plays an important role in soil ecology. You probably don't like it any better than I do when it is necessary to dig a hole to plant a tree, or locate a fencepost. When digging a foundation, you have to put the footings below the frost line to assure that the building won't be heaved out of the soil when the temperature gets below freezing. All that is necessary, of course, because liquid water expands when it freezes. Also, when the soil pores are very cold and water vapor enters the voids, it is attracted to the ice that is already there, making more ice and more soil heaving. That keeps our soils porous so that they have a high infiltration rate—how fast the water enters the soil from the atmosphere and loosens up the subsoil—along with the summer activity of burrowing insects, earthworms, and small mammals. That means it can store more water. Skunks and squirrels make the soil surface open to infiltration, too, as they dig holes in the soil to bury nuts or seek food. But if there were no frost, there probably wouldn't be any soil, and what was there wouldn't have the plants, beasties, and water that together make up a very active and essential part of our environment.

Measuring Soil Moisture

Measuring the amount of water in soil is just about the most difficult measurement in the hydrological cycle. Soil water quantity is important to farmers for irrigation. Scientists need to know how much water is in the soil for plant growth, and how it moves toward a stream or groundwater reservoir. Water supply and flood control reservoir managers measure soil water to forecast rainfall or snowmelt runoff. There are three ways to measure soil water. First, take a soil sample and weigh it, then drive the water out with oven heat and weigh it again. The difference is the amount of water in the soil. But there's a catch: we need to know the water's volume in order to calculate its depth, and thus we have to convert weight to volume. That computation requires information on how accurately the soil and water weights are measured. That is not accurate, and the computed volume measurement is not reliable. A second method measures how much electric current the wet soil can carry, which is more than dry soil. That is not easily related to water volume, however, so the error rate here is also high. A third method involves measuring how many neutrons from a buried radioactive neutron source are deflected by the water molecules in about a cubic foot of soil and returned to a Geiger counter: the number is directly related to soil water volume. This method is the most accurate, but the costs are high. But then, so is the cost of measuring soilwater inaccurately.

Wetlands

Before we called them *wetlands,* they were *swamps*, evil areas by name. Generally considered wastelands, they were there to be drained or filled, resulting in flat land that was highly valued for airports, farms, and industry. Today, we know otherwise: wetlands are vitally important to ecological sustainability, to our very existence. We still spend time, effort, and money restoring, replacing, and protecting wetlands, without being clear about wetland definition or function. Wetlands store water for varying periods of time, providing an environment that is rich in plant and animal biodiversity. Much water chemistry occurs in wetlands, and the ecology of wetlands is complex. Unfortunately, however, many wetland protection laws are not eco*logical.* Some say that wetlands reduce floods and recharge groundwater reservoirs. If you think about it, though, wetlands don't reduce floods. They do provide storage because they are low, flat, and often near streams, but to reduce floods, we would actually need to *drain* wetlands, which would increase their storage capacity. And, wetlands are wet because they are poorly drained and since they don't drain very well, they can't recharge groundwater: if they did, they would be *dry*lands. Wetlands do interact with groundwater: they occur in low spots in the landscape and intersect the water table. Wetlands biodiversity enriches both our environment and our understanding of it. Thus, if we restore habitat, we likely restore wetlands' hydrological functions.

Flushing

Flushing is a natural characteristic of all water bodies except for oceans and non-draining groundwater reservoirs. Thus, upland puddles, wetlands, ponds, lakes, and stream systems all flush nutrients, pollutants, animal wastes, and even carbonic acid from carbon dioxide dissolved in rain and snow downstream, through lakes, floodplain wetlands, and estuaries as the rivers drain to the sea. Materials deposited in the oceans may settle to the bottom, there to become sedimentary rock or to be used by aquatic plants or animals to build tissue to be consumed by other plants or animals, and eventually removed the from the oceans by animals or agents (including humans) that can leave the water. Flushing even occurs in the cells of our and other animals' bodies, where water is used to transport nutrients, gases, and waste products prior to eliminating them from our systems. Thus, the blood stream, coupled with kidneys, digestive system, and liver, concentrates waste products for disposal in liquid and solid wastes, which are flushed down our toilets into the waste treatment plants that discharge treated wastewater into lakes, streams, and rivers. As a consequence, water plays an essential role in the major life processes of respiration, blood chemistry, and digestive processes. Consider this, though: even though a natural function of an environmental water body is to flush wastes, we have a federal law that prohibits dumping wastes into the nation's waters.

Backyard Ponds

A backyard pond can be great fun, but it takes some work to keep it in condition. It is probably best to have two ponds with a waterfall between them, meaning you need a recirculating pump. A small waterfall or fountain keeps the water oxygenated, which is necessary for survival of fish and other pond-dwellers. The ponds need to be at least about eighteen inches deep to provide a haven for aquatic life during the subfreezing winter months. A thermostatic heater that turns on when the temperature reaches forty degrees Fahrenheit keeps the deeper water liquid during cold winters. Pump capacity is measured in the number of cubic feet (or meters) in the ponds: measure or estimate the average depth, width, and length of each pond and multiply their product times 7.5 (the number of gallons in a cubic foot). There are lots of interesting plants that grow in water, and the addition of goldfish, koi, or other species, especially frogs, makes the pond more interesting. One of the major blessings of a pond is that fish eat mosquito eggs, so well-kept ponds control mosquitoes. Another major benefit is to kids, as a miniature world of water. Make sure that they enjoy them safely so they don't slip on or tip rocks at the edge, and perhaps fence the ponds, too. Ponds also attract other animals, including neighborhood dogs, raccoons, possum, skunks, and lots of birds.

Culture and History

Water in Culture

Water makes up such an important and large part of our environment and life that it appears frequently in our thoughts and language. Just think of all the ways in which we use water to mean something common or irregular. Here are some interesting watery cultural expressions: We say that it is "raining cats and dogs." Does that mean pets are as plentiful as raindrops? Two sayings—"It's water over the dam" and "It's water under the bridge"—reflect despair, or our lack of control, or "There's nothing we can do about that!" "Fog so thick that you could cut it with a knife" is a common exaggeration. The belief that "a watched pot never boils" expresses water's high specific heat as well as our impatience. We express being out of one's element by saying "like a fish out of water." A famous western lawyer quipped that "water runs *downhill* in response to gravity, and *uphill* to money." I met another lawyer who reported after studying all interstate stream cases that water runs downhill to lawyers. Water is the subject of classical nursery rhymes as well as popular, classic, folk, and gospel music such as "Singin' in the Rain," "Row, Row, Row Your Boat," and lots more. Of course, we begin our life in water, many of us are baptized in it, and we speak of "washing away our troubles"—or mud. One saying that is definitely not fantasy: "We drink the same water the dinosaurs drank."

Water Is . . .

"Water, water everywhere, nor any drop to drink." That, of course, is a famous quotation from *The Rhyme of the Ancient Mariner* by Samuel Taylor Coleridge, which expresses the frustration of being in the middle of the vast salty ocean but not being able to obtain fresh water. It and many other common sayings reflect a truth about our precious resource. "Don't walk on the water" is a warning about being too demonstrative, pompous, showing power, or some secret. "It's the water" refers to the vital ingredient of a particular brand of beer, and "clear as water" refers to the clean, unpolluted variety that we might observe in natural settings. "It's water under the bridge" or "It's water over the dam" imply that some event or issue—something valuable to us—is gone forever. In the case of water, of course, that is not quite accurate, as water is a renewable resource, that is, different amounts become available at different times. Finally, "We all live downstream" reminds us that whatever we and our upstream neighbors do to the water or the land from which the water flows is detectable in the runoff; in the downstream reaches of a brook, stream, or river; as well as in the groundwater that collects out of sight and mind when our land management practices, flushing toilets, and household wastes eventually turn up in the nearest water body.

The Word Water

Water enters our daily life in ways we don't even think about. I was amazed to discover the many times we use, see, or hear words that mean water or relate to water. For example, the word *water* or words related to water appear in many place names on the land: Watertown, Savannah, Brooklyn, Flushing Meadows, Seneca Falls, Wichita Falls, Rock Springs, Idaho Springs, Grand Forks, Grand Junction, Bay City, and Owen Sound, to mention a dozen. There are also hundreds of names in foreign and Native American languages—or corruptions of them—all over the map: Playa de Ponce, Rio de Janeiro, Hong Kong, Tel Aviv, Le Havre, Wasserburg, and Mzimba. In my fifty-five-year-old *American College Dictionary* by Random House, there are thirty-four definitions of water given in its form as a noun, verb, or adjective. It is the first word in or part of a name for 125 words such as *water buck* and from *waterage* to *watery*. The list includes only one place name, *Waterford*, which is a double water reference and, of course, well known for its expensive crystal. The word *water* derives from Old English *waeter* ("*waiter*"), Icelandic *vatn* ("*watn*"), or Gothic *wato*. In Russian the word is водȧ ("voda"); in German, *wasser*; in Greek, *hydor*; in French, *l'eau*; and in Spanish, *agua*. All sound familiar. *Water* is, indeed, a wondrous word and a universal and essential resource.

Gods and Goddesses

It would take a lot more than one *Water Drop* essay to discuss all history's gods and goddess of water. Here are a few examples of civilization's water deities. Remember, of course, that humans first saw water from the land and had very different views of the oceans, lakes, rivers, and, eventually, the water vapor in the sky that leads to clouds, rain, and snow. Poseidon was the Greek god of the ocean; the Roman god was Neptune. Coventina is the Celtic goddess of rivers, while Latis was the goddess of water and beer. Thor, of course, was the Norse god of sky and thunder. Interestingly, many of the large, inland nations didn't have water deities. Nun is the African god of water and chaos, a fascinating combination that I won't go into here. Hebe was water bearer to the Greek gods. Perhaps one of the most interesting worshippings of the rainfall god was by southwestern Native Americans. After a prolonged drought, they would perform a rain dance. It usually was during the day and involved extensive movement that pounded the already dry soil until it was pulverized. As soil was disturbed, it was lofted and rose on the rising column of hot dry air into the sky where the tiny soil particles served as nuclei on which water vapor accumulates, eventually reaching water-drop size and, more often than one might expect, generating rainfall.

Maxims

Most proverbs or maxims about our environment and weather have a basis in fact. Here are a few. The maxim "Red skies at morning, sailors take warning; red skies at night, sailors' delight" reflects typical movement of weather systems in the northeast United States, where fishing is an important industry. The truism relates the low angle of the rising or setting sun to the direction of the storm system, linked to tomorrow's storminess as systems move to the east. And, as a storm system approaches, the wind will be from the northeast, thus the concern about a nor'easter. Another is "March comes in like a lamb and goes out like a lion," or sometimes vice versa. That's likely because winter weather systems last about seven days. Thus, there are about four and one-half day life cycles for these storm-producing systems. The maxim fits for other months as well, although the cycling is more regular as winter turns to spring, during March. Finally, "If a cat's whiskers droop, it's going to rain; if they're straight out, fair weather." That reflects how animal hair reacts to humidity. In fact, scientific instruments use horsehairs to monitor relative humidity on standard weather instruments.

Nursery Rhymes

How many nursery rhymes and well-known songs do you remember that concern water? I thought there were only a few, and then I started to list them. We are all quite familiar with "Row, Row, Row Your Boat," and of course "Sailing, Sailing, over the Bounding Main" is a song we may have heard long ago. "I Saw a Ship A-Sailing," "It's Raining, It's Pouring," "Michael Row the Boat Ashore," "Mother May I Go Out and Swim?," "My Bonnie Lies over the Ocean," "Rub-a-Dub Dub, Three Men in a Tub," "Itsy Bitsy Spider." And finally, there's "Jack and Jill Went Up the Hill to Fetch a Pail of Water." That one raises some questions, however, because the top of a hill is not a great place to get a large supply of water, unless it is in the form of ice or snow. Maybe they had other things on their minds. There are others, too, and there are several places in nursery rhymes that briefly mention something connected with water. But the ten that I mentioned are probably known to all. There is also a very long list of classical, popular, and folk music that focuses on water. You might have fun with your kids making up a list of water-based songs, symphonies, and other works of musical art. Water plays an important role in our music and language.

Jack and Jill

Folks often ask me whether Jack and Jill were professional hydrologists (water scientists). I have wondered about that myself. After all, if they just wanted clear, cold, pure water, going up the hill was correct. I certainly would have advised them to do that if they wanted better quality. A two-gallon bucket would weigh almost seventeen pounds. But it is a lot easier to carry a bucket of water down than up. Unfortunately, they had difficulty going down, spilling the water so they couldn't use it anyhow. On the other hand, if they knew what they were doing, they were correct in going up the hill. Certainly, if they had wanted a large amount of water, they would have been better off going down the hill, where many streams collect into a river with more water. They probably did want cold, pure water as it comes from a snowfield or glacier: that water would have fewer pollutants dissolved in it, too. It would make great iced tea. Had they wanted warmer water, they would have been better off with river water, although if it came from a groundwater reservoir, it would also be cool, about fifty-five degrees Fahrenheit all year round. Now it may be that they simply wanted to learn about the watershed's distribution of water and carrying a bucket was a ruse. On the other hand, they may have had other things on their minds.

Weather and Climate

Umbrella?

If you were to ask me—a hydrologist—whether you would need an umbrella or snowshoes today, I would want to see current and predicted high- and low-pressure systems; know whether I was in the northern or southern hemisphere, the month, and my latitude. Weather systems generally develop and move to the east. These low-pressure systems are *cyclones*; they bring storms that produce most of the precipitation in the northern United States, initially with warm moist air from the south or southwest; then cool drier winds from the north or northwest. High-pressure systems are *anticyclones*. They usually produce little precipitation, although they can be hot and humid, and may set off thunderstorms. In the northern hemisphere a real rule of thumb is: make a lefthand fist and point the thumb down for low pressure (your fingers—the wind—curl around counterclockwise) and point your thumb up for high pressure (your fingers curl around clockwise). So, if the wind comes from the east or south, a low pressure system is approaching from the west and it is likely to rain first, and then snow if it is cold enough. When a high pressure system approaches, northerly winds will be cool at first, and will then bring warm, probably moist air from the south, without much precipitation. You could also consult the official weather forecast in the paper or on radio, TV, or the Internet.

Chance (1): Rain

When you listen to a weather forecast before going to play golf or to a picnic, you may be confused by the words "chance of rain." For example, "a thirty percent chance of a shower this afternoon" really means that there is only about a one-in-three likelihood of rain, and even then it could be a brief drizzle. Another way to interpret that forecast would be "it will rain on one of three days with weather like this one." Such a forecast might convince me to carry an umbrella, but not necessarily to use it. Actually, weather and precipitation can be forecast pretty accurately, but it is important to listen very carefully to the forecast and its target area: it is easy to be bamboozled. The reason the prediction isn't as accurate as we might like is that lots of different conditions often work together to produce rain or snow. Since those conditions are often cyclical and those cycles don't always coincide, the product—precipitation—is predicted in terms of likelihood of occurrence. And, since storms produce floods, that is true for them, too. You might even hear chance-of-occurrence statements about both precipitation and flooding in the same forecast, and they might not be the same. That is because many things that affect runoff are not included in the list of things that affect rainfall, such as watershed characteristics, when it last rained, and how much moisture is already in the soil.

Full Moon

More often than not, we see the full moon quite clearly. There is a reason for this, and, as you might suspect, it has to do with water. When the sun, earth, and moon line up, strong tides in the oceans and atmosphere are created and the moon is full or new. When the moon pulls the atmosphere up, it creates a dome of high pressure. The air presses down, and there is little movement of moisture-laden air upward for air cooling, condensation, and cloud formation. Pressed down by the high pressure, the air won't reach the dew point, and stays clear. Thus, we usually see the moon on at least one of the several nights when it is full or near-full. Normally, there are long lobes of high pressure radiating from the North Pole that rotate to the east. You can mimic them by spreading your fingers on one hand and setting it on top of a cantaloupe. Each finger represents a high pressure ridge, and the space between them is a low pressure valley. Our weather systems, especially cyclonic storms, originate where the high and low pressure are close enough to create a strong pressure gradient force, one of three forces in the atmosphere that move the air and its water vapor around. The other two are the temperature gradient and the Coriolis force, resulting from Earth's rotation. Together the three help make weather systems.

Humidity

Some people like it dry; some like it humid. The word *humid* means *damp*. So, *humidity* measures air's water vapor content. Actually, there are three different humidity terms: *absolute* and *specific* humidity, used in research and atmospheric physics, and *relative* humidity, used for everyday understanding of the weather and your comfort. The higher the air temperature, the more water vapor it can hold. When the air is less than saturated, the temperature to which the air must be cooled to become saturated is called the *dew point*. The closer the dew point is to the actual air temperature, the more humid it feels. When air at a particular temperature contains as much vapor as it can at that temperature, it is *saturated*. If the temperature of saturated air increases, relative humidity decreases and more vapor could be added. If air is cooled by moving up in the atmosphere, the relative humidity increases and some vapor will have to come out. The water vapor condenses, usually on ice or dust particles if present, or on vegetation forming dew: dew *forms*, it doesn't *fall*. If no particles are present, the air becomes *super-saturated* and drops form very rapidly when dust or ice particles are present. The process is more complicated, too: *evaporation*—literally, *making vapor*—is a cooling process, which means that heat energy is released when condensation occurs. So air cools during condensation, even though heat energy is released in the process.

Clouds

The first definition of the word *cloud* is "a visible mass of vapor, especially one suspended in the sky." But we also use the word *cloud* to mean "to obscure," as when we *cloud* an argument with unclear or ambiguous issues. That use of the word *cloud* has also colored our use of the first definition, and we thus often associate clouds with, as well as being the cause of, bad weather. That's where I thought a *Water Drops* essay was needed. Our derogatory use of the word *cloud* in describing the weather contributes to our negative and even prejudiced view of sunshine obscured by cloud cover and the beneficial precipitation events clouds bring us. Forecast rain and snow storms often herald what weather reporters call "*bad* weather" but, in fact, those precipitation events are ecologically *good*, part of the natural water or hydrological cycle. They return water to Earth after it has been evaporated or transpired by vegetation. During transpiration, water and nutrients are brought up to the growing tips of plants, and evaporation cools the environment where those two processes take place. Clouds cool the Earth, too, by intercepting the Sun's energy and reflecting it to space, keeping us from getting too hot. So, as in Joni Mitchell's wonderful song: "I've looked at clouds from both sides now, from up and down, and still somehow, it's clouds' illusions I recall, I really don't know clouds at all." Do you?

Fog

Being in a fog means you are confused, or you cannot see where you are going—both pretty much the same thing. Fog results when tiny droplets too small to fall develop. They form when the air becomes saturated, that is, when it contains all the water vapor it can hold at the air's dew point temperature. Several things can lower air temperature to the dew point, producing different fog types. First, there is *advective fog*, which is common near water bodies—especially the ocean—when warm dry air flows over cool water and evaporation occurs. When that air then moves over a cooler land surface, or is lifted to cooler altitudes, fog forms. It dissipates when the rising sun warms the air. Second, there is *ground* or *tulé fog*. On a clear night, a flat land surface radiates energy into space, cooling both it and air near the ground, causing it to cool to the dew point. The fog may be only a few feet thick and a foot or so off the ground so you can look for long distances underneath or over it. *Valley fog* occurs when cold heavy air collects in valley bottoms along streams. Lakes may cool humid air that reaches the dew point and becomes *steam fog*. Air above the water body cools to the dew point, and the condensing water vapor appears as little clouds that twist and dance just above the still water surface as local temperature differences create eddies. Fog may persist until after sunrise—when the air warms—so you may be able to get good pictures of steam fog. They could clearly show two essential hydrological processes—*condensation* and *evaporation*. We speak of a "blanket" of fog, implying that it doesn't move, but tulé fog is the only one that is quite stationary: all the others move.

Raindrops

Drops of water are really quite complex. Water's structure the size of drops falling through still air. That rate of fall is called *terminal velocity*. If drops can't go any faster, friction breaks them up, and then those drops reach terminal velocity. If there is a downdraft of air, then the drops can fall faster, and attain bigger sizes. You've probably encountered large raindrops during a thunderstorm, when there are often strong downdrafts of cooler air from upper altitudes. Vegetation, of course, interrupts the fall of raindrops to the forest floor, helping limit soil erosion. The process, which includes snow, is known as *interception*. Interestingly, when small drops fall onto a tree slowly, they may collect at drip points among the branches and larger drops may actually fall to the forest floor. Usually, there is sufficient litter—old leaves, twigs, and nuts or cones—to prevent such large drops from eroding the soil. While it is not a good idea to stand under a tree during a rainstorm, particularly a thunderstorm with potential lightning, you can see large drops forming and falling from drip points, for example through a window, outside your home. They fall irregularly and may concentrate at a few locations, so under a tree is not a good place to locate a rain gauge, much less be with lightning around. But I suspect you knew all that!

Measuring Precipitation

Determining how much water is in rain and snowfall is challenging. Research one hundred years ago resulted in modern raingages. Hydrologists tried long, rectangular troughs for collecting water, but a round opening—with sharp edges to literally split raindrops—is more efficient with less error, so rain and snow gauges are round, eight inches in diameter (so about fifty square inches). In the United States there is about one raingage for each 350 square miles. That's a sample rate of about only one in a billion! Early researchers also explored whether the opening should be parallel to the land surface or perpendicular to the raindrop movement, controllable by wind vanes. Those methods are more complex, so gauges are simply vertical with a horizontal opening. The National Weather Service sets standards on gauge design and specifications. Rainfall is measured in inches or millimeters, not total volume, since we want to know the depth of water at a particular point. Some gauges record timing and the rate of the rainfall, so we can calculate precipitation intensity in depth units per hour. Historical data from many gauges indicate how rainfall varies with time. That important information helps size flood-control structures and stormwater runoff controls. And, there are reliable ways to estimate rainfall between gauges. Snowfall water content is obtained by measuring snow depth and density, yielding inches of water per foot of snow. It is more variable and more difficult to measure.

Measuring Snow

I remember a wonderful cartoon that showed a four-year-old struggling through shoulder-deep snow behind dad. The snow is only up to dad's knees, but he is saying to his offspring, "Don't complain; when I was your age the snow was up to my shoulders!" Extremes of snow are easier to remember, and more difficult to measure, than rainfall. Snowflakes, of course, are solid ice crystals, not liquid water; thus, they are a lot lighter—less dense—than raindrops and hold less water. A rule of thumb is that ten inches of newly fallen snow equals about an inch of liquid water when melted; its density is about 10 percent. So, ten feet of new snowpack contains about twelve inches of water. Over time a snowpack gets thinner and denser, but its water content may stay the same. If the sun shines on the snowpack, or warm winds blow over it, the snow at the surface melts and re-freezes at night, forming a crust, often with higher water content. If the air temperature is near freezing, the pack density may reach 50 percent. At 50 percent water content, a snowpack melts. Snowpacks can also melt from the bottom by soil warming. Next time you see a snowpack, take a shovel and carefully cut away some of it so you can see or even feel the layers of different snow densities from the several storms that contributed to it since the last thaw.

Ice Precipitation

Ice or water falling out of the sky is just another form of *precipitation.* Precipitation occurs when water vapor condenses on dust particles or spontaneously makes ice crystals when the temperature falls below minus forty degrees (which is the same temperature whether you're measuring it Fahrenheit or Centigrade). Dust particles derive from wind, air pollution, or meteor showers. *Condensation* continues until the water drop, snow flake, or ice crystal is heavy enough to fall, or is sent earthward by downdrafts. Different forms of precipitation derive from the sequence of drop, flake, or crystal formation, and air conditions. Thus, *sleet* is rain that freezes falling from the cloud where it formed. *Freezing rain* results when rain drops hit a below-freezing surface. Rain drops freezing within a thunderstorm grow as they circulate in the cloud, finally falling when they get heavy enough; several cycles of this can produce *hail stones* as big as grapefruit. If water vapor condenses directly on surfaces at or below freezing, it forms *black ice*, so identified because there are no air bubbles, making it nearly invisible. *Rime* results when condensation particles fall through super-cooled clouds. It can produce thick ice on any cold surface in the mountains (or airplane wings), or if on tiny snowflakes, the ice particles are called *graupel*, or *snow pellets*. Freezing is a very complex process because while the liquid cools to become a solid, it must give up heat energy, which warms the air in the vicinity of the freezing process.

Black Ice

I sure hope you don't encounter black ice while walking or driving. It is particularly treacherous because you can't see it and it is especially slippery. It only occurs when the air conditions are just right. The air has to be very cold and very moist. Then the *dew point*, the temperature to which the air has to be cooled to reach one hundred percent relative humidity, or *saturation*, is very close to the air temperature. If the air temperature then falls below freezing, the tiniest amount of water on natural and artificial surfaces freezes. Water vapor will then be attracted to that ice and will go directly from a gas to a solid, bypassing the liquid state. That is what makes black ice black: when it forms, there is no chance for air bubbles to dissolve in the ice; normally, bubbles make ice look white because they reflect light. When no bubbles are present, the ice is perfectly clear and is difficult to see, especially at night on pavements such as black asphalt. Since it can't be seen, it is very dangerous. Black ice can develop on vegetation, too: it can be easily identified by its clarity and because it forms equally thick all around the twigs. In contrast, sleet and freezing rain fall downward and then accumulate on top of vegetative parts, so the ice is a lot thicker on the top side.

Morning Ice Crystals

Some really crisp, chilly, and sunny winter morning when you are out and about, skiing, walking the dog, or going to work or school, do you ever notice tiny sparkles in the air? Those are crystals of ice that develop even though there are no clouds. The air near the ground might be below freezing but the temperature a thousand feet up is even colder, at or below minus forty degrees. (You probably noticed that I didn't say "Fahrenheit" or "Centigrade." Actually, it doesn't matter, since minus forty degrees is coincidentally the same in both temperature scales.) At minus forty degrees, water vapor can freeze without having a dust particle or ice crystal to form around and those tiny ice needles develop spontaneously. Those ice crystals won't usually form clouds, but they do fall and are easy to see in the bright sunshine against the blue sky; they look like a shower of sparkles. If you do see clouds by the way, you can tell just by looking at them whether they are ice crystals or liquid water. Ice clouds have fuzzy edges; water clouds have sharp edges. You can see both the next time you see a thunderstorm that builds up into the atmosphere where the temperature gets below freezing. If you are wearing polarizing sunglasses, you can see the difference even more clearly. Ice crystals are one of water's dazzling displays in the cold winter air, one of the many wonders of water that appear before our eyes.

Bad Weather

What goes up comes down; in other words, bad weather storminess is just part of the water cycle. What goes up by evaporation comes down as precipitation in the form of rain, snow, and occasionally various types of fog. The United States gets about thirty inches of precipitation each year. Twenty-two inches evaporates from water bodies or wet surfaces and from plants in the process called *transpiration*. That leaves eight inches to run off the land to the oceans, where it evaporates again. So, the water cycle is indeed circular. Think about this though: a hefty two-thirds of all the precipitation is evaporated or transpired. That's really a lot! We definitely need to pay more attention to that combined *evapotranspiration*, because the eight inches that runs off is a remainder. It is what's left over. If we replace vegetation with paving for our cities and parking lots, the runoff increases, making more stormwater runoff to manage, including floods. What's more, about one-third of all the Sun's energy goes to evaporate and transpire water, and is released in precipitation events, so the energy and water cycles of the Earth are strongly and directly connected: what we do on the land affects local temperature and downstream runoff. Finally, what this means is that all weather is *good* weather: it's all part of life on Earth. There is no *bad* weather: sometimes, it's just that the weather isn't ideal for what *we* want to do.

Storm Types

You have heard weather reporters refer to storms as "lousy weather." But, remember, storms are essential and natural. Life on Earth depends on effects of storms; we would not be here without them. For example, winds and floods play a positive ecological role by removing weak vegetation during destructive storms or by making fresh seedbeds; and floods breach sandbars where a stream enters the ocean, allowing migrating salmon and shad to swim upstream to spawn. Clouds form when the air holds all the water vapor it can, or becomes saturated; then molecules of H_2O combine to form droplets, releasing heat energy. As the droplets in the warmer air rise to regions of colder temperatures, the clouds grow, coalesce, and may make storms. If the upper air levels are dry, the clouds may just disperse. The air cools as it is lifted to cooler levels in the atmosphere by different means, defining three types of storms: *cyclonic*, *convectional*, and *orographic*. Other storms include hurricanes and tornados. Look at your newspaper's weather map, or at the local TV news and weather, or a channel specializing in weather. You can actually watch cyclonic storms develop and move: look for radar or surface weather loops, or watch thunderstorms developing on hot summer days, and orographic storms on the windward side of mountain barriers. There are lots of good books on weather, too, and they make some of the complexities easy to understand.

Storm Types (Again)

I probably discuss rain- and snowstorms in these *Water Drops* essays about as frequently as they occur. They develop under different climate and weather conditions and, often, in different locations on Earth where the conditions for their formation are just right. Storms often follow regularly identified patterns during the traditional calendar year or hydrological seasons, or by location, and many may be particularly destructive. Cyclonic, frontal, or air mass storms are cyclical. They produce most of the annual precipitation in the northern United States. Convectional storms are typical of hot, relatively dry climates that prevail in the southwestern United States. Although storms differ by name, storm precipitation always develops from the same fundamental process: that is, some mechanism—air masses, solar heating, or mountains—lifts warm, moist air to cooler levels in the atmosphere, where there is sufficient cooling so that the water vapor condenses into clouds of solid flakes, hail, or liquid drops. Lake effect storms are considered orographic storms that provide heavy annual precipitation to local regions, and many lake effect events also involve lifting water vapor over mountain barriers to produce snow far removed from the originating lake. Most importantly, storms are the primary means whereby water is delivered from Earth's atmosphere to Earth's surface, the driving process underlying the hydrological cycle. In sum, remember that storms are a natural part of our environment and—as long as they are not unreasonably destructive—are good weather.

Cyclonic Storms

You can make a miniature model of a low-pressure system by swirling water around in a pot. Most of the precipitation in the northern United States is from similar systems, called *cyclonic storms*. These are low-pressure weather systems that spin counterclockwise and start along the polar front, where cool, dry polar air is deflected to the west by Earth's spin, while warm, moist tropical air is deflected to the east. As it rotates, the cyclonic storm intensity increases and moves eastward. On the east or leading side of the low, a warm front develops as the tropical warm moist air is lifted over the colder air. Its movement is slowed as the cold air is trapped under the overriding warmer air and a warm front forms between the two air masses; rainfall is usually gentle and spread over a wide area. Along the trailing edge of the low, polar air wedges under the less-dense warmer air, which readily moves upward, creating a more violent cold front on the ground. Precipitation is more intense, is over a smaller area, and is of short duration. Thunderstorms are more likely along a cold front. Cyclones typically dissipate after five to seven days, having matured when the cold front catches up to the warm front, producing a *stationary front*. Cyclones move about seven hundred miles a day in winter, and are slower and less frequent in summer. You can watch cyclones' life cycles on Internet radar maps.

Convectional Storms

Most of us are scared when we first hear thunder, and are fascinated by the flashes of lightning that produced it. I like the sound so much it accompanies booting up my computer. During the summer months and often all year in the warmer southern parts of the United States, most of the rainfall comes from *convectional storms*. These storms usually form in the afternoon, when Earth's surface has been heated by the sun. That, in turn, heats the air, causing it to rise, where it gradually cools. The rising air forms the familiar puffy fair-weather *cumulus* clouds, but as they continue to build up, threatening *cumulonimbus* clouds known as *thunderheads* will form. While most rainfall is generated within about fifteen hundred feet of the ground, thunderstorms may top out at one hundred thousand feet. These storms can produce dangerous and destructive lightning, thunder, hail, and even tornados. Intense thunderstorms usually cover a small area, and are rarely more than ten miles in diameter, but they can be larger. Thunderstorms may also develop along a cold front, and look like a bunch of soldiers along what is often called a *squall line*. They may also develop in association with hurricanes, or over mountains as orographic storms.The clouds and behavior look pretty much the same as convectional thunderstorms since the process is the same for all types. Occasionally, thunderstorms may develop along a warm front in winter as thunder snow.

Orographic Storms

The rain in Spain stays mostly in the mountains. So does the snow. Mountains are barriers to wind. As air flows over them, it cools at higher altitudes. Cooler air can't hold as much water vapor, so the rising air cools to the *dew point*, the air temperature at which the relative humidity reaches 100 percent (saturation) and clouds form, often producing precipitation. These are *orographic storms*, which produce most of the rain and snow on windward coastal mountains. Annual precipitation on the Olympic Peninsula reaches two hundred inches or more, almost seven times the nation's average. Rocky Mountain snowfall is typically twenty feet deep. Windward shores of tropical islands also have orographic precipitation: Mauna Loa in Hawaii gets an average of 450 inches. As the drier air descends on the lee side of mountains, it becomes warmer because heat was released when water vapor condensed, because temperatures are higher at lower elevations, and because the air is squeezed as lower elevation pressure increases; hot, dry winds result, called *foehns, Chinook,* and *Santa Ana* winds. They quickly melt and evaporate snow, dry soils, and spread wildfires. Cold, dry air crossing large lakes produces considerable amounts of *lake effect* precipitation, which may include orographic effects. Annual snowfall on the Tug Hill Plateau east of Lake Ontario amounts to an average of three hundred inches, while the cities of Albany, Buffalo, Rochester, and Syracuse compete for heaviest annual snowfall.

Lake Effect

Living in the vicinity of the Great Lakes, or just driving through the area, can be a big challenge because of three words—*lake effect snow.* Lake effect snow occurs near other lakes, too. During winter, winds driven by polar air generate narrow bands of snow downwind of Lakes Erie and Ontario when they are free of ice. Those Canadian winter winds are cold and dry. As they flow across the lake they are warmed by the water and pick up moisture by evaporation since the relative humidity of the air is quite low. Then, as the air passes over the downwind shore, it is cooled by the colder land surface and/or by snow already on the ground. The air can also be lifted over hills or mountains like the Tug Hill Plateau, generating orographic precipitation. The resulting cooling produces the dense cloud bands that generate extensive and often very heavy snowfall in very localized areas. When Doppler radar was installed in the Great Lakes region, it was discovered that lake effect snow and its orographic effect produced a considerable proportion of the winter snowfall in the Adirondack and Catskill Mountains in New York and even in Pennsylvania's Pocono Mountains. If you watch the radar loops on the Internet you will see that lake effect snow may even extend into the Berkshire Mountains in Massachusetts. And, remember: lake effect snow runoff doesn't add to the water level of Lake Ontario: that's where the water came from in the first place!

Hurricanes and Tornados

Hurricanes and *tornados* are particularly destructive events that develop from extreme differences in wind, water, and temperature. Tornados are very intense small diameter whirlwinds spawned by convectional storms, or thunderstorm clouds. Winds can exceed two hundred miles an hour. Tornados result from strong contrasts of temperature and moisture. There are excellent accounts of tornados at Internet sites, and in educational films and weather books. Spring tornado outbreaks are likely when a mass of Arctic dry air extends down over the central plains from Texas and Oklahoma to Mississippi and Georgia, and meets tropical moist air from the Gulf of Mexico. The violent front that develops can spawn many tornados, and the region is known as Tornado Alley. It may extend all the way into the Ohio River valley. Hurricanes develop in the tropical Atlantic from June to December. The same type of storm in the tropical Pacific is a *typhoon*. Hurricanes shift our Sun's heat energy from the tropics poleward, and when these intense storms impinge on land they cause considerable damage from wind and flooding, caused by torrential rain and the storm surge. This surge is the result of the hurricane's very low pressure, which allows the sea level to rise at the center or eye of the storm. Also, the winds in front of the hurricane push the deeper water up and over the near-shore shallows. Whether hurricane or tornado, take heed of any warnings you receive because both can be very deadly

Drought

By the time you feel thirsty, you've probably been without water too long. You become aware when you are thirsty. The same is true for *drought.* When a flood starts, you know it right away. One can easily identify a flood: it is bank-full capacity. But it is difficult to define drought, or point to the time when one starts. Droughts develop slowly over time, and it is hard to know when to say, "We are starting to have a drought." Just like floods, droughts are natural, though, and the climate and weather cycles reflect that. In fact, floods and droughts are really part of the same spectrum of streamflows, from high to low. There are actually three types of drought: all three are described as having less than the normal or average amount of water in some time period, usually a season, but sometimes a year, or several years. First, if there is too little precipitation, it is known as *meteorological* drought. This, of course, occurs when the annual or seasonal precipitation is below the long-term average for one or more years. A second type of drought is *agricultural* drought, signaled by wilting vegetation. Third, there is *hydrological* drought, when streamflow falls below the long-term average. Obviously, there is likely to be meteorological drought before agricultural or hydrological drought, but it is still very difficult to identify when a drought begins and, consequently, when and how to start water conservation measures.

Meteorological Drought

One of the three types of drought is precipitation—or *meteorological*—drought. Since it is almost impossible to know when a drought begins, one has to wait until water is scarce. Even with the sophisticated Palmer Drought Index, one must wait until it is clear that precipitation has been below normal for some time to know that a drought exists. And, a drought on a city watershed doesn't mean that it is dry in the city. In 1981, New York's Department of Environmental Conservation postponed a drought warning to New York City residents because, although there was a serious deficiency of water in the Catskill watersheds, it was raining in the city. The public would not have bought the idea that water was in short supply. The department waited until there were several days of rain-free weather before imposing drought restrictions. Since precipitation depends upon a large number of influencing factors, such as temperature, time of year and sunlight, atmospheric pressure, and the evaporation rate that returns water to the atmosphere, dry periods occur for different lengths of time and may or may not continue, and drought is thus difficult to plan for. Meteorological drought is particularly difficult for farmers and ranchers, and for all who depend on surface water supply systems. Since managing drought is so difficult, it probably wil be best for us to eliminate our waste of water.

Agricultural Drought

Defining drought sounds simple but it's not. *Agricultural* drought is evaluated in terms of crop growth, demands for water, and the amount of moisture in the soil. The National Weather Service uses the Palmer Drought Index to evaluate this type of drought. Unfortunately, it takes time to know when a drought has really started, and last month's records to know whether a drought is really occurring. Calculations consume time, and by the time a drought is evident, it may be too late to do anything about it. Also, some crops may be able to survive with very little water whereas other crops show effects of drought after short periods of low rainfall or drying soils. Since many crop irrigation systems get water from streamflows, low precipitation and runoff—*hydrological* drought—may be the cause of agricultural drought. More likely, lengthy periods between rainfall and hot, low-humidity weather may trigger an agricultural drought. Usually agricultural drought is associated with both hydrological (low river-flow) and meteorological (low rainfall) drought. A weekly drought index is better than a daily (too short) or monthly (too long) interval, and should be based on the amount of soil moisture, amount of loss from soil and plants, and rainfall. One must also consider the amount of water in man-made storage available for irrigation. Farmers use sunflowers to indicate the need for irrigation: tall wilted sunflowers are highly visible and intolerant of low soil moisture, signaling the need for irrigation.

Hydrological Drought

Droughts may last for years and can extend over wide areas, causing severe hardship for communities, homeowners, farmers, and ranchers. City water-supply managers report *hydrological* drought as low river-flows or reservoir levels. Hydrological drought results from a deficiency in precipitation and runoff—meteorological drought—and the actual definition, or specific amount or percentage of normal flow varies from one location to another. Both hydrological and meteorological drought types are at the other end of the natural spectrum of flood peaks and occur especially when temperatures are warm and water leaves the soil by vegetative transpiration or by evaporation before it can run off. In addition to not having enough water for humans or animals to drink, there may be insufficient water for growing crops, from which farmers make a livelihood and ranchers feed their stock. Even if precipitation and/or river flow is low, there may be ample water in groundwater storage. The term *drought* applies to low flows of water in normally temperate regions: deserts are *always* droughty. Just as flood frequency and magnitude may be evaluated from streamflow records, so too can drought. Floods, however, are associated with varying flow values; drought is evaluated as anything below normal, with a limit, of course, of zero. Probability data are particularly useful for helping make decisions about whether or not there is enough water in a stream or river during dry years to develop a supply for a municipality or industry.

Climate Change Terminology

Weather, *climate*, and *global warming* are three terms that describe different characteristics of our atmosphere, that thin envelope of gas (including water vapor) that encloses planet Earth. The three terms vary by area and time. Weather is the sum of conditions for a relatively short time and a relatively limited area. For example, the day might be hot and humid, cold and snowing, or windy but clear. Climate is the sum of weather over a longer period and of an area or region larger than just the local weather site. The time involved could be a season, or a month of winter, spring, or summer. We also refer to climate to characterize a locale, such as a tropical island. Global warming is the long-range and wide-area change that appears to be occurring now. I say *appears* because the temperature could change to *cooling* instead of *warming*. Paradoxically, warming could actually start a new Ice Age. As Earth warms from an interglacial period—such as we are now in—enough cloud cover could develop that blocks enough sunshine to cool Earth's surface and lead to more snow and ice. We have barely a century of reliable temperature, rain and snowfall, humidity, wind speed and direction, and air pressure records. Scientists infer longer period information from drilling ice cores in Greenland and Antarctica. Thus, weather represents our environmental experience associated with days, climate with years, and global warming with experience over lifetimes.

Canary

Water is so intimately associated with life that it might be considered as our canary in a coal mine. If something is wrong with the water resource, something is wrong—or needs fixing—with *us*. Water underlies *all* aspects of our lives. It delivers our nutrients, removes wastes, regulates our temperature, and quenches our thirst. It underlies external environments, notably weather and climate; grows food, flowers, and forests; provides transport, cooling, and fire control; and is the focus of arts and architecture. So, as we find water resources changing and identify underlying environmental problems, we need to pay attention. At present, melting mountain glaciers and the polar and Greenland ice caps, rising ocean levels, increases in the number and severity and/or change in start, length, or end of season of hurricanes and tornados, and rising temperatures are all symptoms of excess carbon dioxide in the atmosphere. CO_2 is chemically and physically connected to water and water vapor. Water is greatly involved in the development of thunderstorms: their lightning sets off forest and grassland fires, positive feedback that adds to the atmosphere's temperature and carbon dioxide levels and is detrimental to a livable environment. It may be offset by long-term changes in Earth's climate that humankind may or may not survive. Certainly, the civilization we have built will not survive unimpaired by such changes. We humans are mobile and upright yet at the mercy of conditions that we adversely affect primarily by our conversion of inorganic carbon to carbon dioxide, and by our sheer numbers.

Our Greenhouse

Water plays many important roles in controlling Earth's atmosphere and temperature, our planetary greenhouse. For example, carbon dioxide in the atmosphere readily dissolves in water, making rain (and snow) slightly acidic as carbonic acid, thus reducing the amount of carbon dioxide in the atmosphere and its greenhouse effect. Methane, discharged from dead organic matter in wetlands and directly from animals, is another important greenhouse gas. Water itself plays a vital role in how some of the carbon dioxide gets to the atmosphere in the first place, since the breathing in of oxygen and breathing out of carbon dioxide takes place in the presence of water in animals' lungs. The reverse, of course, takes place in plants, where carbon dioxide is inhaled and oxygen is exhaled by the plant cells where water is again available for the conversion. Water vapor is also directly important in the atmosphere, since it is the most potent of greenhouse gases because it traps longwave radiation from the Earth, keeping some of it from escaping to space. As experts have noted: "Without water vapor, the most important atmospheric greenhouse gas, the planet's surface temperature would be well below freezing."* Water is very slow to change its temperature, which means it has a very high specific heat, requiring lots of calories to evaporate. Finally, water vapor condenses into clouds and reflects incoming sunlight back to space, preventing the Sun's energy from warming the planet's surface.

---

*Gaffen, D. J., R. J. Ross, and J. Gille (2000). Water vapor's role in climate revisited at follow-up Chapman Conference, *Eos Trans. AGU*, 81(15), 158 doi:10.1029/00EO00109.

Global Warming

The Earth has been both a lot warmer and a lot colder than it is today. Right now we are definitely getting warmer based on direct measurements started about a century ago and on inferences about gases, pollen grains, and other contents of the two-mile-thick ice sheets of Greenland and Antarctica. We are in an interglacial period now. In the middle of the last century, it was predicted that our atmosphere would have to warm up before a new Ice Age would start: increased temperature would mean more clouds, shielding the planet from the Sun, thus making it colder. Since the start of the Industrial Revolution, atmospheric carbon dioxide concentration has been steadily—and, recently, rapidly—increasing. CO_2 is a fragile atmospheric regulator that helps control Earth's temperature by the *greenhouse effect.* Atmospheric CO_2 increases from forest fires and burning fossil fuels; its current level is higher than at any time in the past three hundred million years. Most of it is absorbed by water and plants. Currently, the warmer air dramatically warms the oceans, resulting in more, more frequent, and more severe storms, which normally move equatorial heat energy received from our Sun toward the poles. Earth is definitely warming, although not yet as much as it did in the Carboniferous Age, when excess carbon in the atmosphere was deposited as plants and animals died, becoming today's fossil fuels. Water plays essential and critical roles in the temperature regulation of our planet.

Hydrology

Hydrology

Hydrology is the study of water in both the natural and disturbed environments. Probably the oldest correctly recorded hydrological observer was the author of Ecclesiastes, who wrote, "The rivers run into the sea, yet the sea is not full." Leonardo da Vinci and Bernard Pallisy in the fifteenth and sixteenth centuries were important observers of nature, but even da Vinci made some mistakes and may have been the source of misinformation that the oceans produced springs by pressure. Scientific hydrology really began with Pierre Perrault, who with Edmé Mariotté published *On the Origin of Springs* in 1678. He determined there was sufficient precipitation on the Seine River watershed to supply a year's runoff, thus defining the hydrological cycle. Robert Horton wrote about the relationship between infiltration and runoff in the 1930s, and C. W. Thornthwaite created a numerical water balance in the 1940s. Scientists, engineers, and courts continually define more specific hydrological processes. New measurement instruments and techniques are added every day. One new field is isotope hydrology, which evaluates water sources and circulation times in the landscape. Most attention has probably been given to the hydrology of the world's temperate zones, but there is critical need for water resources management in tropical and arid zones, too. Arid- and tropical-zone hydrology may be defined in terms of the relative amounts of annual precipitation and potential evaporation and transpiration, or evapotranspiration. The relationship may be expressed in an aridity or humidity index.

Hydrology (Second Definition)

When did it last rain? That's important. The first definition of the word *hydrology* is the study of water in both natural and disturbed environments. Another definition describes the relevant condition of the water resource at a particular time. This is especially important when an extreme event such as a heavy rainstorm threatens. Also called *antecedent moisture conditions*, the watershed after a long drought would be dry and not likely to produce a flood, whereas if the soil were full from earlier rains or snowmelt, or the natural storage capacity of the soil would be low because of compaction, and a rain might produce excessive flooding. If watershed forests have been harvested or killed by insects or disease, the soil can be quite wet because normal evaporation and transpiration has not occurred, and the soil reservoir is full. Any added water could produce a flood. The term *hydrology* is also useful in describing what a particular stream has been doing when a storm occurs, or a drought is underway. Water-quality sampling also needs to have information on the condition of the water resource—the hydrology—at sampling time. This is important because at the start of a runoff event, whatever is in the stream is first diluted and then rises to a peak value before returning to the pre-runoff-event level. Thus, it is crucial to know the hydrologic character of the stream at the time the sample was taken.

Annual Hydrograph

A *hydrograph* charts water flow in a stream over time. It is a graph of streamwater flow for an hour, a day, a series of years, or an average year. It represents the stream. During an entire year, streamflow reflects changing watershed conditions. During a summer thunderstorm, the hydrograph rises rapidly as water falls in the channel and runs through the soil near the stream or from any impervious surfaces, such as roads, roofs, or lawns. That type of runoff—storm runoff—doesn't last very long. Water that does sink into the soil may join the groundwater reservoir, often a large body of water saturating deeper soil and fractured rock. That water flows to the stream slowly, and on a large watershed, dominates the streamflow. That is, most of the yearly runoff is from the groundwater reservoir. That runoff has different properties, too. It is usually around fifty-five degrees Fahrenheit, so it is useful for cooling. If it has been in sandy soil, it will have silica dissolved in it; if the rock is limestone, the water will be alkaline; if the tree cover is pine, the water may be acid. Groundwater runoff rises and falls slowly, so on a large watershed, the change in runoff rate is slower than from stormwater runoff. It is particularly important to understand the stormwater and groundwater runoff behavior when we start to manage the watersheds' many resources.

Hydrological Seasons

The water—or hydrological—year has three major processes or seasons: *precipitation*, *evaporative losses*, and *runoff*. Storage isn't included because it is roughly the same at the beginning and end of each year. Evaporation and transpiration—the movement of water through vegetation from the soil and out the leaves—is called *evapotranspiration*. It is greatest during the summer, utilizing soil moisture to satisfy the warm air's thirst for water vapor. That is why the soil is driest around October first, the start of the hydrological year. So the first hydrological season is *soil moisture recharge,* starting around October 1 and lasting until there has been enough precipitation to recharge the soil moisture. If precipitation continues—or if there is snowmelt—it goes to the next season, the season of *maximum groundwater recharge* or *maximum runoff*, depending on where the water goes. That is generally the period we call *spring*. Of course: that's when the water that flows through the soil begins to appear as springs. As temperatures warm up for summer and plants put out new growth, the season of *maximum evapotranspiration* (or *maximum soil moisture utilization*) begins, ending with autumn's leaf fall to start the cycle over again. In some very dry regions, there may be no annual runoff (although local storms may produce small amounts of rainfall that don't enter the soil and that run off as stormwater runoff), so there are only two hydrological seasons.

Happy New Year

In October, you can say *Happy New Year!* because it is around the first of October that the water (or hydrological) year officially begins. (In the southern hemisphere, the start of the water year would be at the beginning of April: see if you can figure out why.) The water year is more of a natural year than our calendar year. The soil is dry because summer rainfall is usually low and long hours of sunshine and warm temperatures in June, July, and August increase evaporation (literally "making vapor'); thus, vegetation removes water from the soil through the roots, stems, branches, and leaves, in a process that is called *transpiration*. Transpiration brings nutrients from the soil up to the wood and leaves, keeps the vegetation moist and flexible, and cools the air. When the monthly rainfall—and in winter, snowfall—is greater than monthly evaporation and transpiration, the soil is recharged in early fall. This is the first season of the water year, the *season of soil moisture recharge*. As the soil storage capacity fills, water can flow down to groundwater and laterally to streams. In the second season, the *season of maximum runoff*, the creeks, streams, and rivers spring forth. That's why we call that calendar season *spring*. The third hydrological season, then, is the *season of maximum evaporation and transpiration* (or *soil moisture utilization*). Knowing about water behavior is useful to understand how to manage our water. So, Happy New Year! Happy New *Water* Year, that is.

Season of Soil Moisture Recharge

The hydrological or water year starts with the *season of soil moisture recharge*. It is first because we identify the water or hydrological year by the twelve months that start October 1. At that time, after the long hot summer, the moisture in the soil has been withdrawn by the plants that pump water to the atmosphere in the process called *transpiration*. That is an evaporative process and the two processes—evaporation and transpiration—are combined into one word: *evapotranspiration*. Some vegetative root systems may dry out soils to great depths. The dry soils are refilled with water from fall rains or winter snowmelt. The water is drawn into the soil's tiny pores called *capillary pores* or *micropores*. The larger *noncapillary* or *macropores* can't draw the water in since the large pores are more open and don't pull as much. The pulling is called *tension:* capillary pores exhibit very high tensions. Strong upward tension by vegetation overcomes the force of gravity that acts to pull soil water down. Some trees are ten times as high as the thirty-two foot height to which normal atmospheric pressure of about fourteen pounds per square inch pushes water up. The trees, grass, and agricultural crops send lots of water up into the atmosphere after the soil is recharged, but by that time it is winter, and evapotranspiration starts when our summer season begins. In the southern hemisphere, of course, the months are shifted by half the year.

Season of Maximum Runoff

The middle season of the hydrological year is the *season of maximum runoff*. It starts when the preceding season of soil moisture recharge is over and water flows through the soil to streams and/or groundwater reservoirs. It ends when the air temperatures warm up, causing water to evaporate from water bodies and the vegetation to grow and transpire water. The runoff is made up of three parts: water flowing out of the groundwater reservoir, called *groundwater runoff* or *base flow;* water flowing through the soil called *subsurface flow*; and *surface runoff*, which includes water flowing off of impervious surfaces and water that falls directly in stream channels. These three types of flow or runoff behave differently, although the surface and subsurface runoff often behave more like each other than like base flow. Base flow supplies water to the stream or river most of the year, especially on large watersheds, where its pattern of runoff dominates the annual runoff. On small watersheds, there is less groundwater storage, so the annual pattern is dominated by surface runoff with more ups and downs and is called *flashy*. Different patterns of watershed runoff establish a unique pattern of stream ecology: how the aquatic environment varies, providing protection, dissolved gases, and food for fish and other aquatic species. If we harvest trees or other vegetation, or cover the soil's surface with pavements, for example, we change that runoff pattern, often adversely affecting a stream's ecology.

Season of Maximum Evapotranspiration

The summer season is vacation time for most, but for the water around us it's time to do important work. Hydrologists refer to the *season of maximum evapotranspiration* or *maximum soil moisture utilization.* It occurs when warm summer temperatures pull water into the atmosphere. Water moving up through plants brings nutrients that help grow new wood, leaves, and fruit. And when it evaporates, it cools the air. The amount of evapotranspiration—or ET—depends on the amount of energy available. More hours of daylight mean more energy, as does the higher sun angle. Wind is also important: as ET occurs, it blows away the moist air near the leaf surface. Water departs vegetation through microscopic openings usually on the bottom of leaves, called *stomates* or *stomata.* Two guard cells control the stomata by reacting to the air's relative humidity, preventing the plant from drying out. Vegetation that grows near water, such as wetland trees and shrubs, can pump tremendous amounts of water into the atmosphere. So can some trees and grasses that grow roots down to great depths—sometimes forty feet or more—to where there is plenty of water. Vegetative species living in or near water are called *phreatophytes*, literally "water-loving plants." There are lots of invasive phreatophytes, such as salt cedar, which seriously decrease western states' streamflow. You can also use a willow tree to help dry out your lawn if it is too wet. Willows are also phreatophytes.

Arid Zone Hydrology

You must think I'm pretty dumb to talk about water in the desert, so I won't, but there *is* actually a lot of water *beneath* many deserts. It is so deep, however, that it costs too much to pump it up, and even more to get it where it is needed. Also, once it's on the surface, it could evaporate quickly. Arid zones are best described by the relationship between annual precipitation and potential evaporation and transpiration (evapotranspiration). An area is arid if the year's possible evapotranspiration is greater than the year's precipitation. The timing of the two processes is also important: some areas may be arid for part of the year and humid at other times. Our own southwest region is pretty arid: many areas get only a few inches of rain each year, and the possible evaporation is much greater. The soil can get so dry between scattered storms that the water can't even get into the soil. Many of these areas are also low in elevation, and rivers, often fed by snowmelt in nearby mountains, run through them. The population in many of our arid zones is growing rapidly as people move to these so-called sunbelts. And they need lots of water there! Water is needed to keep cool, and to irrigate agricultural crops, lawns, and gardens, and for fighting fires. It is going to cost a lot of money to supply these arid zones with water.

Deserts

Water is the stuff of life but many people have moved to where the water *isn't*, like the southwestern United States. Recent research shows that some of the world's biggest deserts, such as the Sahara, have *lots* of water, but it is so deep that it is not worth pumping up. Once that water reaches the surface, it quickly evaporates. The southwestern United States is very arid, with many areas receiving only a few inches of precipitation each year. Annual evaporation potential far exceeds rainfall. Many areas in the southwest are relatively low in elevation, and the rivers that run through them dry up because of the low water flow, high heat, and overuse of the river's precious water. In western Arizona, the Colorado River carries lots of snowmelt from the Rocky Mountains, but that water also carries sediment because the watershed soils are not well protected by desert vegetation. Plus, the water has been used for irrigation agriculture and it dissolves soil salts, making the water unfit to drink or use for much else. There's very little rainfall runoff into these rivers because the soil gets so dry between very infrequent storms that the water evaporates or rapidly runs off as unusable floods before sinking into the soil. Yet despite being an arid zone, Arizona and Nevada are some of the fastest growing areas in the United States. You can imagine that it is going to cost a lot of money to keep supplying these arid zones with water.

Tropical Hydrology

Living in the tropics would certainly be a challenge to this temperate-zone-loving hydrologist! Tropical regions are defined by year-round or seasonal warm temperatures or by having more than enough annual precipitation to meet the heat's demand for evaporation and transpiration. Tropics are typically hot and humid, and they have dense vegetation dominated by vines, tall trees, and air plants that can remove water from humid air. The prolonged rainfall period is called the monsoon season. Cherapungi, in India, often has monsoons combined with hurricanes and extreme mountain rainfall, up to a thousand inches per year, and often in only six months! Tropical hydrology certainly differs from temperate zone hydrology. So, tropical water resource management must provide ample storage during a prolonged dry season, and limitations on irrigation of agricultural crops may be severe. Second, trade winds produce excess localized rainfall on the windward coast. Third, tropical forests are particularly important since good forest cover must be maintained, usually by masses of roots that hold the soil and minimize erosion. Many tropical countries experience severe flooding and loss of topsoil after deforestation. Results include complete loss of resource base and great loss of life, property, and crops. High economic costs for restoration follow.

Cold Regions Hydrology

Earth's cold regions—mountain glaciers and polar ice caps—are important because they contain two-thirds of Earth's fresh water; they directly affect the great southern and northern oceans. Unusual characteristics and measurement difficulties limit our understanding, and observation networks of snowfall, snow depth, ice extent, soil frost, and streamflow are imperfect. Surfaces of vegetation and ice or snow have solar radiation reflection capabilities very different from temperate zone regions, and local energy balances are readily upset, as is now happening as our climate warms. Hydrology of cold regions exhibit low energy input, affecting snow and ice storage, and melt. Many Arctic, Antarctic, and alpine regions are covered with tundra, plant life dominated by short, cold growing seasons and frozen soil. Tree vegetation is scarce or absent. Observing snow accumulation, redistribution, and melt; water content of frozen soils and peat bogs, and ice-covered streamflow provide considerable hydrological uncertainty. Glaciers and polar ice caps are melting at an accelerated rate, affecting ocean water temperature and salinity, further impacting ocean circulation, feedback loops, surface and subsurface vegetation, and wildlife populations. Cold regions occupy so *much* area and hold so *much* water that we need to pay a lot more attention to them.

Runoff

Water runs off land surfaces for a variety of reasons. If the surface is impervious, all the water that falls on it runs off except for a usually small amount that evaporates if the surface is warm, and a similar amount that evaporates after the rainfall or snowmelt stops; that's the amount of water necessary to wet the surface. Similarly, water that infiltrates the soil—moves out of the atmosphere into the lithosphere—first replaces empty storage space formed by earlier drainage to deeper levels, such as the groundwater reservoir or nearby streams, or by transpiration of soil water through the vegetation that is dependent on vegetation and atmospheric characteristics. If the runoff water goes to the groundwater reservoir, it usually moves out quite slowly and is called *base flow*. Water that flows through the soil to streams joins water that falls directly into the stream channels, and surface runoff and is called *storm flow*. These two types of flow, base and stormwater, make up our yearly runoff. A graph of the stormwater runoff over time is called the storm hydrograph, while the combined storm and base flows make up the annual hydrograph. Not only do they look different, the water quality of each is different since water is a universal solvent and picks up many of the substances it contacts. So, water chemistry helps us determine where water came from as well as whether it is stormwater or base flow.

Measuring Runoff

Watching a stream always makes me wonder, How much water is flowing past me? Simple math provides the answer, and the results may astound you. You need a reasonably straight and uniform stretch or reach of stream, about one hundred feet long, although fifty feet will do. Measure or pace off that length, and then measure or estimate the width and depth of the water in feet or meters. Time how long it takes for some leaves or twigs you throw into the stream to run the reach. Divide the reach length by the number of seconds recorded to determine the water velocity in feet per second. Multiply the width times the depth to get the cross-section area in square feet, and then multiply by the velocity to get the streamflow—or discharge—in cubic feet or meters per second. For example, if some leaves take thirty seconds to run a one hundred foot reach, the velocity is about three feet per second. If the reach averages two feet deep by five feet wide, the cross-sectional area is ten square feet. Multiply that times the velocity to get about thirty-three cubic feet per second or almost three million cubic feet per day, more than you can drink. At seven and one-half gallons per cubic foot, that equals enough personal use water for a town of 140,000 people for a day.

Groundwater Runoff

Groundwater is out of sight, but not out of mind. When there is no rain or snowmelt to make a storm hydrograph, water may still be flowing gradually out of the groundwater reservoir: it is called *base flow*. Base flow responds more slowly than storm flow because it is sluggish and less accessible to the stream. Also, evaporation and transpiration by vegetation usually don't decrease water in the groundwater reservoir. From the large natural reservoir, groundwater runoff increases during or after rainfall and snowmelt events; it lasts a long time, and reaches a minimum in the fall when soil water replenishment is low and summer evapotranspiration has been high. Groundwater runoff dominates the annual pattern of runoff on large watersheds; small watershed runoff is mostly storm flow. On very large watersheds almost all runoff is from groundwater. Sometimes base and storm flow may be differentiated by temperature and chemistry. Groundwater runoff is usually around fifty-five degrees Fahrenheit year round; if the water is a lot colder or warmer, it is storm flow. If storm water has contacted sandy soil, it will have silica dissolved in it. If underlying rocks are limestone, then the groundwater will be alkaline, since calcium carbonate in the rock dissolves readily in water. Next time you see a stream see if you can figure out whether it is storm- or base flow by knowing the month, the date of the last rain or snowmelt, and watershed size.

Stormwater Runoff

Stream water during and immediately after a storm or rapid snowmelt event is called stormwater runoff. Think about where rainfall or snowmelt runoff went before your street was built, how it got there, and where it goes now. Before the suburban or urban area was built, rain and snowmelt water soaked into the soil and flowed out slowly. Plant roots, winter frost, and burrowing animals and insects kept the soil porous; vegetation pumped some of that water out of the soil to the atmosphere, while the rest of the water flowed to rivers or to groundwater. Stormwater runoff combines with the natural features of the watershed—the geology, topography, soil, and vegetation—to create a unique storm runoff pattern for your stream. Since urbanization changes several of the characteristics, the runoff pattern also changes. With most of the area impervious, water can't get into the soil or groundwater reservoirs, and local flooding occurs. When it does get downstream, it causes downstream flooding. And, of course, water flowing over the soil surface picks up and carries soil particles, or sediment. When the stream floods and spreads out downstream, it deposits the sediment as unwelcome mud. The runoff water also carries lots of pollutants, including lawn fertilizer, fungicides and insecticides, pet droppings, parking lot and highway marking paint, salt, and petroleum products. And, loss of water to groundwater reservoirs reduces the structure of the soil leading to a sinking land surface, or subsidence.

Storm Hydrograph

Storm runoff can be very dangerous: it rises quickly and can stall your car or be over your head before you know it. Storm runoff floods streams, often carrying lots of sediment. During a rain storm or snowmelt event, the watershed-draining stream rises rapidly, reaches a high point or flood peak, and then decreases more slowly. The chart of storm runoff over time is called the storm hydrograph. The other type of streamflow is runoff from groundwater, water from below the water table. It rises and falls more slowly than storm flow. Storm hydrograph shape and timing are determined by storm size, type, and movement, and by watershed geology, topography, soil characteristics such as depth and texture, type of vegetation and land use, and by time since the last storm: a unique pattern for each watershed. If the features change, so does the storm runoff. Building a subdivision changes the watershed and, therefore, the shape and timing of the storm hydrograph. As much as 60 to 70 percent of a suburb may be impervious streets, sidewalks, patios, and driveways, compacted lawns, and house roofs: all make rainfall and snowmelt water run off more rapidly causing local and downstream flooding. With more pervious natural areas, the water would soak into the soil, from where it is taken up by plants, or flows slowly to a nearby stream or down to groundwater.

Storm Flow

When I was a kid in New York City, I used to play engineer and build dams in the street gutters to control the snowmelt runoff, and then break them to watch the flood. That's my first recollection of runoff. Storm flow is also called quick flow, or event runoff. It occurs during a rainfall or snowmelt event, and is separate from base flow, which is water that flows more slowly from below the water table. Storm flow may consist of runoff from land uses that inhibit the natural movement of water into the soil. Such surfaces include impervious street and sidewalk pavements, roofs, compacted lawns and driveways, and where grazing, agricultural equipment, or even recreation activities compact the soil surface. Storm flow also includes water that flows rapidly through a porous soil and goes directly to a local stream instead of to the water table and groundwater. It includes water that falls directly into the stream or into lakes or wetlands connected directly to the draining stream. The storm flow period may be only slightly longer than the storm, and the storm runoff is also known as stormwater. Stormwater runoff is now regulated by law because it may contain sediments, animal droppings, and lawn and agricultural fertilizers, herbicides, and pesticides that degrade water quality. Even though this land use runoff is a nonpoint source of pollution, it now requires a discharge permit like other point sources that have more stringent legal controls.

Stormwater Runoff Quality

Many people think of stormwater runoff as an annoyance, a temporary flooding or lawn erosion problem. The quality of runoff from urban and suburban areas is poor, primarily because the water does not filter through the soil. Stormwater collects on pavements and runs to storm sewers that deliver the runoff directly to local streams, and sometimes to waste treatment plants. Stormwater runoff doesn't need as much treatment as sanitary sewage, but it can still be pretty nasty. It may contain pet wastes, lawn fertilizer, oil, salt from snow control, and even toxic paints from parking lots and highway markings. The Clean Water Act requires new storm and sanitary sewers to be separate and old systems to be modified. The federal Environmental Protect Agency regulates (through the states) and conducts educational programs where regulation is not easily accomplished. For example, the EPA recommends ten ways to improve stormwater runoff quality. These include: Use fertilizer sparingly and sweep up driveways and walks; never dump anything down storm drains; vegetate bare spots in the yard; compost lawn and garden wastes; avoid using pesticides, herbicides, and rodenticides; direct downspouts away from paved areas; use a carwash instead of the driveway; recycle motor oil and check car for leaks; pick up after your pet; and, if you have a septic tank, inspect and clean it regularly. Other information about the EPA and water quality can be found at its Web site, epa.gov.

Floods

Floods are considered natural disasters, but Mother Nature would say that floods are just part of a watershed's normal behavior. Floods are normal occurrences: the other end of the spectrum is drought. Floods, of course, occur when the storage capacity of the watershed is full and runs over, just as a glass of water runs over the top if too much water is poured in. On small watersheds, thunderstorms are typically the cause of floods. Their rate of rainfall—the rainfall intensity—often exceeds the infiltration capacity—the ability of the soil to absorb it—and the excess runs off to the nearest low place, a dry depression, or a running rivulet, stream, or river. On large watersheds, a rapid regional warmup in spring provides snowmelt; the water from the melting snow and ice runs off quickly. Floods on medium-sized watersheds typically result from hurricanes or monsoons, which deliver large quantities of rainfall in short periods. To prevent floods, natural storage, such as in the soil, has to be protected along with the maintenance of high soil infiltration rates. They may be enhanced by aerating the soil surface with a piercing tool on your lawn. Floodplains and wetlands are a river's natural storage areas, so when we build on those lands, or build levees to contain floods, we don't change the amount of water that the watershed produces: we only restrict where it can go, making the flood even higher.

Infiltration

Floods can be dangerous. They may be caused when water doesn't enter the soil fast enough and runs off the surface. Normally, the movement of the water into the soil is faster than the rate of rainfall to the soil. The rate of rainfall delivery to the ground surface is called the *rainfall intensity*. The rate the ground surface can take in water is called the infiltration capacity. Both are reported in inches or millimeters per hour. If the rainfall intensity exceeds the infiltration capacity, then there is surface runoff and, in all likelihood, flooding. Infiltration capacity is higher for the big pore spaces of a sandy soil than for the tiny pore spaces of a clay soil and, of course, the infiltration capacity of concrete or asphalt is zero. A mixture of soil particle sizes is usually called a *loam*; it has a high infiltration capacity along with the ability to hold a fairly large amount of water for plant growth. Under most natural—that is, undisturbed—conditions, the soil's infiltration capacity is greater than the rainfall intensity, and all the water enters the soil. Shallow soils and steep mountain slopes often combine with high rainfall intensities to produce surface runoff, But if the soil surface has been compacted or changed to material with a lower infiltration capacity, then water runs off the surface and causes local and downstream flooding as well as a lot of sediment in the stream.

Chance (2): Floods

Floods can be terrifying. It doesn't matter whether the flood or peak flow is two or ten feet deep: what matters is whether the water is over your nose. Floods are caused by storm runoff or snowmelt that could occur any time. Hydrologists readily analyze a stream's peak flows for as little as twenty years to predict how big a flood occurs once in a hundred years, but not when it will actually occur. During constant climate, we expect the one-hundred-year flood to be observed once in any one-hundred-year period, the one percent flood. Just as tossing a coin has the same likelihood of producing heads or tails each toss, the chance of observing the one-hundred-year flood is the same every year. But, there is also a chance of observing the one-hundred-*and-one*-year, the one-hundred-*and-two*-year floods and the *thousand*-year flood each year, all greater than the one-hundred-year-flood. A much better statement would be "the chance of having the one-hundred-year flood *equaled or exceeded*." That's actually two and a half times in any one-hundred-year period, about once every forty years, the result of dividing one hundred by two and a half. The one hundred year flood is usually the design flood for flood control projects, but it only protects up to that amount, not over it. Watershed land use changes increase flood peaks, requiring new analysis, and changing the design flood.

More Flooding

We measure flood peaks and how often they occur to identify the floodplain maps that determine homeowners' and businesses' eligibility for flood insurance. Floodplains are the natural storage place for excess runoff, and it is a good idea to stay out of them. We don't. You can only get flood insurance if you are located above the one-hundred-year-flood limit—the design flood. Flood peak records tell hydrologists how frequently floods occur, but not when. We define the one-hundred-year flood as that size observed once in a hundred years; but the less likely one-hundred-and-one-year flood, and the thousand-year flood could also occur. There are floods of other magnitudes, too. So, it is better to say that the one-hundred-year flood may be expected to be equaled or exceeded more than once in one hundred years, in fact once in about forty years. The flood frequency curve is a characteristic of a river. The longer we keep flood records, the greater the flood we will likely observe. It is also true there will be greater floods as we change the watershed's ability to absorb rainfall and snowmelt, because that changes the flood frequency curve. And the flood frequency curve changes as our climate changes, too. So, we can expect to see more and greater floods as time marches on. Count on it.

Floodplains

Geographer Gilbert White spent much of his life trying to get folks to understand that floods were natural events. He pleaded that we should be careful about how we manipulate them and the lands and waters in and around them because, as he said, "the floodplain belongs to the river." As of this writing, the Cedar River in Iowa is experiencing a five-hundred-year flood. That is not quite the correct way to express how severe that flow of water is. The proper way to state it is that the size of the flood can be expected to be equaled or exceeded at that place on the river once in five hundred years. That is a lot different from saying you're going to observe it once in five hundred years. There is also the chance of observing the five-hundred-*and-one*-year flood, and the five-hundred-*and-two*-year flood, and the *six-hundred*-year flood. You get the idea. If you think that once you have experienced such a flood that you are safe for another five hundred years, forget it. You could have the five-hundred-year-flood exceeded next year. Several years ago, Grand Forks had a major flood. They'll have a lot more too, as will any urban area with *lake, pond, brook, river, forks, falls,* or *ocean view* as a part of their name. I'd be reluctant to live in any of them. Wouldn't you?

Flood Insurance

I wouldn't live in a floodplain any more than I would live on the slope of a volcano. Insurance companies that write flood insurance policies may have to make flood insurance available to you. Flood insurance has been available since the 1968 Flood Insurance Act. The federal government subsidizes flood insurance because while fires may be isolated to one or a few homes, floods affect everyone below the flood's water level. Insurance companies would be out of business. The terms include that your house is above the one-hundred-year flood level for the particular area, as determined and mapped by the Corps of Engineers, and that everyone who lives in the floodplain is treated the same way. That is to avoid the local zoning agency's being sued for a regulatory taking. With current increase in frequency, intensity, and amount of precipitation due to climate change, your likelihood of being flooded out is increasing. To avoid bankruptcy, flood insurance companies may still have to offer flood insurance, but the premiums may be so high that they are two or three times the value of the house. So, if you live in, or are contemplating buying property in an urban area with place names that include *rapids, falls, forks, lake, pond, river,* even *view,* I would think again. You are getting a warning message: your chances of being flooded may be ultra high. Look elsewhere.

Three Rivers, Three Floods

I had the dubious experience of living through a major flood. In December 1964, the Eel River in California crested at 950,000 cubic feet per second—or cfs—a bit less than one million. I was there. That's a lot of water! It would fill a typical domed stadium in fifty seconds. But it is only slightly under the 1.1 million cfs record flood on the Columbia River at Portland, Oregon, in 1945, the same year the Mississippi River at Re River Landing, Arkansas, peaked at about 1.2 million cfs. What is very interesting about these three record flood peaks for the twentieth century is the very different watershed sizes. The Mississippi River watershed is about three hundred thousand square miles. The Columbia River is one-tenth that: thirty thousand square miles. And the Eel River watershed is only three thousand square miles. Yet all three streams produced approximately the same peak flow. On the Mississippi, the flow developed from snowmelt over a wide area, and flood peak was only about three cubic feet per second (cfs) per square mile; the Columbia peak flow, after a rainy fall, freeze, heavy winter snows, and rapid spring warmup, was about thirty cfs per square mile. But on the Eel River, fifty inches of rain in nine days produced three hundred cfs per square mile. That's the way the water flows; floods.

Economics, Management, and Policy

Using Water

Each of us uses a lot of water: to keep our bodies cool and moist and to remove wastes, we drink four or five pints each day. That may include coffee, tea, orange juice, or beer. It takes about 3 gallons of water to brew a pint of beer. It takes 50 gallons to make a glass of orange juice; 280 for a Sunday newspaper; 800 to grow a pound of hamburger; and 100,000 gallons to produce an automobile. Every time we flush the toilet, 3 gallons go down the drain, unless it's a low flush toilet using half that. Daily household water use totals over 100 gallons per person, not including water to keep our lawns green, wash the car, hose off sidewalks and patios, or wash windows, clothes, and dishes. Nationwide, and for all purposes, we withdraw more than *four hundred billion* gallons per day, three-fourths of which is surface water. Over 80 percent is used for irrigation and power production. Most of our appliances these days are pretty efficient in using water, but we can use them more efficiently, for example by waiting until the dishwasher is full. We can't really save water, but we can reduce the amount we use, leaving more water for others. There are lots of people in the world who have only a fraction of the water we have in the United States, as little as a quart, and it's often dirty.

Water in the World

We are fortunate to be able to enjoy a glass of orange juice reconstituted with safe water in the morning, and to have excess water to wash ourselves, our clothes, and our property! About 1.2 billion people don't have access to clean drinking water, and about 45 percent of the world's population doesn't have access to basic sanitation services. The impact of a lack of clean water appears in countries where less than 20 percent of the population has access to clean water: the death rate is two hundred per thousand for children under five. In many regions groundwater is pumped faster than the annual rain and snowmelt can replenish it, and underground supplies for current and future populations are not sustainable. Mining groundwater causes regions to sink, settle, or subside, jeopardizing roads, houses, and utilities and, ultimately, the economy of the region. It is not sustainable. Desalinizing ocean water is not normally feasible because it still requires expensive pumping to get it where it is needed. And, the thousands of dams that have been built severely disrupt the natural characteristics of rivers, lakes, and streams to the extent that our biological world is severely impaired. You can find out about some of these problems and some potential ways to fix Earth's water problems by putting "water problems" into an Internet search engine. There are also lots of organizations working on different geographic scales to improve the sustainability of our life-giving water.

Storage

More than 99 percent of all Earth's water is in storage. About 97 percent is in the oceans, too salty to drink. Another 2 percent is in ice, diminishing by global warming. Of the 1 percent remaining, three-fourths is in groundwater, and nearly all the rest is in lakes, leaving less than 1 percent in motion between Earth's atmosphere and land. A particular molecule of water could change from stored water to water in motion between two storage sites. That water would exhibit characteristics of the storage it was just in. So, water flowing out of sandy soil will have silica in it; water flowing out of groundwater is about fifty-five degrees Fahrenheit; and water flowing from a wetland exhibits high acidity and low silica. That's because rotting plants remove oxygen from and release carbon dioxide to the water, and because growing plants use silica to build cell walls. As humans grow in population, we need more and more water to assure our health, cleanliness, and survival. Thus, we build artificial water storage, especially dams and reservoirs, to provide continuous supplies of water during periods of low or no rainfall or during low runoff seasons. Because the water to which we actually have access is such a small portion of the total, we have to consider the water in storage in planning and management.

Resource Buffers

Water is distributed on Earth in a lopsided manner, with about 97 percent of it in the oceans, two-thirds of the remainder in ice, three-quarters of the rest in groundwater, and almost all the rest is in lakes. One-fifth of that is in one lake, Lake Baikal. It turns out that all our natural resources exhibit that lopsidedness: even the mass of the Sun, of the largest and the three largest planets. It is true of space, energy, perhaps time, and even the simplest atom, hydrogen: its single proton is two thousand times the mass of the single electron. With everything made of atoms, that kind of distribution is quite logical. We learned that lopsidedness first in biology, where it takes lots of pollen grains or sperm to fertilize one egg; where lots of seedlings germinate, but only a few survive, and so on. That lopsidedness applies in our culture, too: most of Earth's people live in a few urban areas, within a short distance of the coast, and only a small percentage hold most of the wealth; three water companies now control 70 percent of the world's private systems. We use only a small percentage of our brain cells, and so on. The larger resource amount that we don't use directly is a buffer, and we need to protect these vitally important buffers—like all that water in the oceans—because our sustainability depends on their health.

Watersheds

A recent educational publication on watershed planning that did not include a definition of the word *watershed* prompted the thought of my using the word too casually. There is a lesson in there about the failure of professionals to let folks know what we are talking about. Mea culpa. A watershed is a natural or disturbed unit of land on which water collects, is stored, and runs off to a common outlet, such as a stream, lake, or ocean. That is a simple definition, so we need to be more specific. For planning purposes, we need to ask, "What is the difference between large and small watersheds?" If we really want to help those who make decisions about watershed planning and management, the answer must be functional, not physical size–based. Doctor Ven Te Chow, author of the *Handbook of Hydrology,* in 1964 suggested a functional definition differentiating between large and small watersheds. He recommended that the difference between the two should be whether stormwater or base flow dominated the annual runoff. Stormwater runoff includes water that falls directly in stream channels, flows over the surface directly to the stream, or infiltrates the soil near the stream and quickly flows to it. Base flow, on the other hand, comes from the groundwater reservoir and runs off much more slowly. This difference is important since we should make watershed management decisions based on which type of runoff we wish to control for our use or protection.

Water Balance

Water is the watershed's asset, just as money is your bank account's asset. Both change monthly; you expect your bank balance to grow; but the water balance in any location is about the same from year to year, changing with weather and climatic cycles. A watershed's water balance shows inflow and outflow and the change in storage, just like your bank balance. The amount of water is reported in inches (or millimeters) of depth. Your annual bank balance in dollars is your deposits minus your expenditures on January first. The annual water balance in depth is precipitation minus evapotranspiration and runoff, calculated when the amount of water in soil storage is lowest, at the start of the water year, which runs from October first to September thirtieth. Evapotranspiration is the combined evaporation from surfaces that interrupts precipitation and from puddles, lakes, and streams, and transpiration is the water released through tiny openings in leaves, called stomates. Any change in the balance indicates a climate variation. Annual numbers vary greatly from deserts to tropical environments; the average annual United States water balance components are about thirty inches of precipitation, twenty-two inches of evapotranspiration, and eight inches of runoff. In dry regions, additional water may have to be used to irrigate croplands. And, if the watershed's land use changes, the water balance will also change, usually with less water going into storage and more to runoff and flooding—in other words, the storage assets are lost.

Watershed Research

How watersheds work is investigated around the world. Some of the earliest research on watersheds was in South Africa and England. Two research watersheds were set up by the Forest Service more than a century ago, on the upper Rio Grande watershed at Wagon Wheel Gap, Colorado, and at San Dimas, near Los Angeles. When I got out of college I worked at the Coweeta Hydrologic Laboratory in the North Carolina Mountains. Coweeta is a big watershed with about fifty small, gauged watersheds used to study effects of land use on streamflow and water quality, and how runoff reacts to removing forest cover completely or in strips, or just cutting underbrush. Studies were also conducted to find out how to do such studies, as at San Dimas, such as how to place rain and stream gauges. The Hubbard Brook watersheds in New Hampshire were used to investigate runoff water quality starting in 1955. The Forest Service operates experimental watersheds around the country, and the Agricultural Research Service operates in farming areas, and watershed research at Coshocton, Ohio. Many universities also have research watersheds and laboratories. All the research watersheds have intensive gauge networks of rain, groundwater and soil moisture gauges, some experiments on how and how much water moves through plants and, of course, stream gauges to record the timing and amount of runoff. Many welcome visitors, too, where you can see experiments and talk with the research personnel.

Watershed Functions

I found that a lot of people weren't sure what a watershed does or how it functions. A watershed is an area of land on which water collects (the British call them catchments) and runs off (we call them watersheds), so there are two functions: collection and runoff. Between collection and runoff, water is stored. Thus, watersheds collect, store, and discharge water. The discharge is shown by storm and annual hydrographs that show runoff rate over time. Whatever comes in contact with water causes chemical changes to occur, which affects water quality and provides habitat for plants and animals. Also, storm runoff peaks decrease—are attenuated—as the peak moves downstream, and water quality changes as chemicals are flushed from the system. Those are the seven watershed functions: collection, storage, discharge, chemical reactions, habitat, attenuation, and flushing. Natural and manmade changes in climate, weather, geological history, local topography, vegetation type and its management, soils, and characteristics of the channel all affect watershed functions and runoff behavior. For example, pavement affects whether, when, and how fast water enters soil storage, which in turn affects chemistry and runoff water quality. If we understand watershed functions, we can manage our watersheds to provide storage and runoff when, where, and how we want them.

Wetland Watersheds

Wetlands breed mosquitoes, and I overreact to their bites. However, Wetlands are very important features in the landscape: how watershed functions and wetland characteristics interact is important, too. Relevant coinciding wetland and watershed functions include storage and discharge of water, habitat, and water chemistry. Watershed behavior is impacted by both the presence and location of a wetland (or lake). Most important is the proximity of a wetland to the outlet of its watershed. Type and extent of the runoff-causing event, existing soil moisture and groundwater levels, and timing of events are critical factors in wetland-watershed interactions. A wetland watershed is usually difficult to identify and/or delineate, indicative of its poor drainage. Generally, wetlands reduce high and increase low streamflows. Thus, wetland characteristics are evident in the watershed's storm, seasonal and annual discharge patterns, and modify the watershed's hydrological functions. Wetland evidence also shows up in stream chemistry: wetlands are sinks for dissolved silica—required for plant cell walls—and excess vegetative decomposition demands oxygen, releasing carbon dioxide. The presence of wetlands may prolong floods by their presence near the stream but does not reduce them, nor do wetlands recharge groundwater. Wetlands are particularly important for their biodiversity.

Watershed Planning

Since we all live on a watershed of one size or another, we all have the opportunity to be informed about and active in watershed planning. Watershed plans take a variety of diferent forms, and may be prepared by individuals, groups, consultants, agencies, or nongovernment organizations (NGOs). Formalized watershed planning process books and manuals cover what to do and how to do it. For example, watershed planning for farms involves several preformed steps or tiers, starting with Tier One, a questionnaire about what the problems and challenges are, land managing units, goals, and a broad overview. Tier Two links the problems and goals to plan development—how the goals might be achieved. Tier Three includes the design and engineering of potential Best Management Practices to realize the goals, as well as budget, funding sources, and a time line. Tier Four covers implementation. And Tier Five reports and follows up on the overall plan. This Whole Farm Planning process started with New York City's Catskill Watershed planning, and was formalized by the Natural Resources Conservation Service. Watershed planning commences with an inventory covering all the factors affecting the watershed's water resources, including climate, soil, vegetation, land use patterns, ownership, transportation networks, and current and anticipated trends in population numbers and distribution.

Watershed Management

Watershed management is about a century old. The late-1800s' research at Colorado showed that watershed vegetation, soil, and runoff are closely related. Watershed management considers all natural resources as a system related to water. That means focusing resource planning on the drainage basin of a stream, river, wetland, or estuary. It means making decisions about how the forest or grassland vegetation, soils, steep slopes, flat floodplains, and wildlife lands are used by us. We now know how changes to the landscape affect the water that falls on, and runs through and off, the land by what we do. Watershed management is the planned manipulation of natural or disturbed drainage features to effect a desired change in or maintain a desired water resource condition. For example, to get the most water from a watershed it would be best to cut all the vegetation and pave it. On some tropical islands, humans have done that to ensure good water supplies for themselves. But there are bad effects of that, too: the water can run off very quickly because the vegetation no longer protects the soil or transpires water back to the atmosphere. We would have to build storage for any excess water. We know how to cut forests to increase snow storage, and even delay its melting until we need it during a warm summer. A century of research has shown us how we can manage our watersheds.

Objectives of Watershed Management

To manage natural resources effectively, it is necessary to practice ecosystem management. From my standpoint as a hydrologist, that means watershed management, the planned manipulation of one or more features of a natural or disturbed environment to effect a desired change in or maintain a desired condition of the water resource. There are three general objectives of watershed management policies and practices. First is rehabilitation of abandoned, abused, or even naturally altered lands that produce excess sediment, polluted runoff, or ill-timed runoff itself. The second objective is protection, especially for normal and sensitive areas or activities that might lead to the need for rehabilitation measures. Historically, these two objectives seem backward, but the need for rehabilitation was first identified by George Perkins Marsh, who in 1874 wrote the classic *The Earth as Modified by Human Action*. The third objective is enhancement, for we know from a century of research that we can effectively manipulate vegetative cover and runoff to improve functions on the watershed. Thus, we can maintain a young, rapidly growing forest that uses water faster and thus maintains more of a soil storage reservoir to help reduce floods. Or we can cut small areas of forest cover sequentially to manipulate snow accumulation and melt, thereby reducing spring floods. Practical application of Best Management Practices are particularly important means for working with the watershed to ensure regular delivery of high quality water while making productive use of the landscape.

Municipal Water Supplies

Where do you get your water? City water comes from groundwater, oceans, lakes, rivers, or watersheds. Each has pros and cons. Groundwater wells require expensive pumping, and water may be contaminated by percolating wastes from septic systems, irrigated crops, or urban runoff. Desalinizing ocean water is expensive, and it must be pumped up for pressurized delivery. Lakes vary in size, and waters may be polluted by urban runoff from the city for which the water is supplied. Chicago gets water from huge Lake Michigan, using more lake water to flush wastes down the Chicago River to the Mississippi. Syracuse gets its water from Lakes Ontario and Skaneateles. Many smaller communities get gravity water from nearby high lakes. Denver, east of the Continental Divide, gets water through gravity tunnels from reservoirs west of the Divide. Portland, Oregon, gets water from forested watersheds on Mount Hood, and San Francisco from the Don Pedro Reservoir on the Tuolumne River in the Hetch Hetchy valley, site of the first major battle in the United States over conservation in the early 1900s. Most of New York City water comes from two thousand square miles of watersheds in the Catskill Mountains tunneled under the Hudson River. And Boston gets its water from forested watersheds of lakes created by dams in the Quabbin Valley. The American Water Works Association provides water treatment operator training and professional certification, assuring safe drinking water for all. Check out the source of *your* water.

Forested Water Supplies

Seattle, Washington, gets its water from land it manages for its high value timber, keeping the forest healthy and the water quality high. A forester manages lands of the 140 square mile, now city-owned Cedar River watershed, covered with Douglas fir, cedar, and spruce. It was closed to logging in the 1990s. If forest lands are closed to harvesting, normal debris, and blown-down vegetation build up to levels that are susceptible to fire and provide attractive hosts to disease or insects. Keeping the forest healthy is good protection for the precious water resource. Two other cities with forested watersheds are Segovia in Spain and Melbourne, Australia. Segovia, sitting on a three-thousand-foot-high plateau northwest of Madrid, has a two-thousand-year-old still operating ten-mile-long Roman aqueduct discharging into a natural gravel layer tapped by city dwellers' wells. Segovia also has a newer modern dam and reservoir nearby, storing water from a managed forested watershed. Denver's western slope watersheds provide water by gravity through a twenty-six-mile tunnel under the Continental Divide. The high-elevation forest and alpine lands are managed by the Forest Service. Risk of fire prompted Melbourne to close its forest lands, excluding any private use, a policy that may lead to even more fires, watershed damage, and poor water quality. Carefully managed natural cover and alert managers provide a valuable first line of defense against disasters while yielding excellent water.

Woods and Water

Woods and water. The connection is fundamental to ecology and sustainability. In the 1700s and 1800s forests were harvested for fuel, construction, transportation, and furniture industries. Forest lands were cut over and abandoned as the timber barons moved westward in search of new forest resources. Major shifts in forest land management policy reflected changing awareness concerning the consequences of continued exploitation without consideration for the watershed values of forest lands. The Pittsburgh and Johnstown floods led to research into the relationship between forest cover and water yield. The fire induced by the San Francisco earthquake drew attention to water supplies adequate for fire fighting and expanding urban areas. These conflicted with the wilderness movement, thus influencing relationships between the Park Service and Forest Service and led to establishment of New York's Adirondack Park, in response to growing citizen concerns with wilderness preservation. One of the significant consequences of changes in forest policy was that forest growth in the 1950s exceeded the annual cut as sound forest management began to dominate our federal and commercial forest lands. Notable actors in this long history included Oliver Wendell Holmes, the Roosevelts, Pinchot, Olmstead, Muir, Marshall, Bennett, and Leopold. I highly recommend the biographies of these individuals, for they all played major roles in the nation's history, in understanding our natural resources, and in our continuing need for both wood and water.

Quantity, Quality, and Regimen

The most important things to know about your water supply are first, how much is there? Second, what is its quality? And third, how is it distributed in time, or simply, when do we get it and how much during different time periods? That last property of a water supply is called regimen, the same word applied to how often and how much medication you take. The quality and quantity are—of course—rather straightforward properties of water. Anyone developing a water well, for example, needs to know the water's quantity, quality, and regimen to properly assess the ability of the water supply to meet the needs of a single family's groundwater well or a community's massive water supply. Professional well drillers are licensed and are usually required to provide their drilling logs to the state, as well as quantity, quality, and regimen information to the person paying the bill for the well. They also have a professional organization—the National Groundwater Association—providing conferences and publications to help members and anyone associated with the groundwater industry—do a better and better job.The quality question is rather simply addressed because we can ask, "Is it tasty and safe to drink?" The American Water Works Association provides certified workshops and classes as well as certified water (and waste) treatment plant operators. They work closely with the federal Environmental Protection Agency and local and state health departments concerning water quality.

Measuring Water Quality

Measuring water quality is tricky. Regulators want to know how much—or load—of a pollutant there is to control polluters. That requires measuring the concentration, since a stream's load is the concentration per unit of flow times the flow rate. The content per unit is of primary interest to research modelers. Since concentration of many pollutants depends on how fast the water is moving, and water doesn't move uniformly across the stream profile, several measurements need to be made. That is quite costly, because water quality samples and laboratory tests are expensive. Also, the velocity of stream water varies across the stream as well as with its depth. In a fairly uniform channel, the velocity of the water is greatest near the middle of the channel and about one-third of the way from the stream's top to its bottom. But measuring water quality concentration only there doesn't reflect the total load because the concentration may be much lower elsewhere. And you have probably watched a stream and seen water eddying and even flowing upstream near the edges. Even if the quantity of pollutants and quality of streamflow are measured, there is also the problem of measuring the pollutants during a period of time, most importantly during a stormwater runoff event when most of the year's pollutants are flushed from the watershed. Thus, to know the true water quality load, concentrations, different stream locations, and different times must all be sampled.

Water Quality Laws

Congress has only recently been involved in controlling water quality. It started with the Corps of Engineers in 1899 when Congress passed the Refuse Act, which assigned a permitting process for dumping in navigable waters. The limitation to navigable waters is to ensure the constitutionality of the act, a federal responsibility as stated in the Constitution. The meaning of the word *dumping* was not clearly defined; it was aimed at hard fill, but has been interpreted as anything that degrades water quality. At mid-century, the public clamored for better federal water quality legislation, resulting in the 1948 Water Pollution Control Act, which sidestepped public concern about water pollution by leaving water quality control to the states. A comprehensive water Quality Act was enacted in 1965 amending the 1948 act, gradually tightening water quality control, providing tax incentives for industrial cleanup, and requiring the federal government to set water quality standards. When President Nixon's executive order established the Environmental Protection Agency (EPA) in 1970, it included the former Public Health Service, which had had responsibility for those water quality standards. Finally, detailed controls and federal-state cooperation were applied to specific definitions of point and nonpoint sources of water pollution in the 1972 Water Quality Amendments, which became known as the Clean Water Act after further amendments in 1977. Major amendments in 1987 and 2002 further strengthened the legislation.

Drinking Water

You may have gotten a notice from your public water supplier about a quality problem. Hopefully, you won't, but every once in a while a water main break or flood or drought may result in a letter, or a radio or TV message, that describes a water problem. For example, an unusually heavy rainfall may have washed sediment into the water supply system. The notice might suggest boiling the water just to be sure it's safe to drink. That is because the capability of the water supply system to remove sediment and all the accompanying microorganisms might not be completely satisfied by existing filters. Water supply personnel will notify you when the water is again safe. They may also inform water users with general water quality information: how much chlorine it contains to control infectious organisms, or whether fluorine is added to help control youngsters' tooth decay. There are also some old disease-carrying organisms, including dangerous ones such as diphtheria and typhoid, and new ones such as giardia and cryptosporidium. The Environmental Protection Agency is responsible for regulating public water supply quality. The American Water Works Association provides water supply technicians with education and up-to-date information on testing sampling and procedures to meet licensing standards. The EPA provides a Safe Drinking Water Hotline and works through state and local health departments to ensure safe water supplies. That is not true for bottled water, however, although bottled water may come from municipal taps.

Water Meters

Where is your water meter? Do you know how it is read? Like a gasoline pump, water meters count the number of units of water used in each household so the water supplier can properly bill the homeowner, keeping wasteful water use habits, and leaky faucets, down. A unit is usually one hundred cubic feet; about seven and one-half gallons. So just multiply the number of units used by seven and one-half. Until recently, most meters were mechanical, mounted on the basement water supply pipe and read periodically by a meter reader. The supply pipe comes into the house below soil freezing depth, and the mechanical meter was inside. Many water utilities are now using high tech meters that are read by telephone or radio signals, to eliminate the time-consuming job and intrusion of going inside someone's home. Some water department computers automatically telephone the meter at night. If the line is or becomes busy with a voice call, it tries again later. Nowadays, a meter reader points a handheld radio device at the meter to get the reading, which is later downloaded to the water department's computers for billing. In older apartment buildings, identically laid-out apartments receive water from vertical pipe systems instead of horizontally to each rented apartment, making water delivery to all the bathrooms and kitchens more efficient, but not attributable to each apartment. It would be difficult to determine each owner's usage without expensive pipe rerouting.

Pricing Water

Most of us don't know what we pay for water. Well, we do know how much we pay for bottled water: a lot! At $1.75 per quart, that's seven dollars a gallon, *double* what we pay for gasoline. When we buy water from a utility, such as a municipal supply system, the rate is usually given per unit, with a unit being one hundred cubic feet. Since a cubic foot contains about seven and one-half gallons, the unit contains 748 gallons. That's enough for about five people using a typical 150 gallons per person per day. Water billing is usually done quarterly, but it can be monthly to encourage water conservation. To reduce wasting water, the amount paid per unit increases after the first unit. In really dry areas—like the sunbelts—water bills can skyrocket, especially if you're not careful how much you use. One southern California bill was $1.09 for the first four units used; then tripled for the next four, then six, then four of the units used again. Under that billing scheme, one month's bill was $194. That sure discourages lawn watering, car washing, and lengthy showers! Here in the humid east, our typical quarterly bill is thirty to fifty dollars, or around twenty cents per thousand gallons. Domestic water is pretty cheap, and is generally subsidized to assure sufficient water for health purposes.

Bottled Water

Bottled water is a ten billion dollar per year business. More than three hundred million Americans annually average about twenty-four gallons of water. A gallon of bottled water may cost as much as ten thousand times as much as a gallon of municipal tap water; and may even *be* tap water! I remember during a sunrise stroll at a fancy hotel seeing an employee filling bottles from a garden hose. It is also having damaging effects on the environment: using petroleum products, shipping energy pollution, and in landfills. Bottling businesses may compete for natural water resources, affecting the water price from traditional municipal wells, watersheds, and desalinization plants. Some groundwater and spring fed water sources in Florida developed by bottled water companies ship the bottles out of state in return for special development consideration. That displaces residents and keeps would-be residents from local land and water. Federal, state, and local agencies regularly test public water supplies for substances that might sicken us. Untested bottled water is great to have for victims of natural disasters where local water supplies have compromised. It costs a great deal to transport, however, and the bottles themselves may be environmentally damaging, in landfills as well as in their manufacture. I must admit to having bottled water when I am traveling and nervous about local water quality. Otherwise, I prefer the well or watershed drinking water sources, and I am satisfied that they are quite safe.

Hot Shower

Think about this the next time you are taking a long hot shower. I just did. About fourteen thousand BTUs of natural gas energy are used to heat the twenty gallons of cold municipal water for a typical shower; more in my case. That energy likely came from sunlight ten million years ago that was used by a plant to grow leaves and branches. The vegetation may have been eaten by a wooly mammoth that was killed and eaten by a Tyrannosaurus Rex that later died in a swamp. There it remained for perhaps another million years, decomposing and becoming one with the organic material from plants, omnivores, bugs, and inhabitants of the swamp. Eventually, the whole mess of other lifeless and once-living debris sank deeper, was covered by eroded soil sediments and maybe by an ocean, then perhaps rock, for several thousand years. The pressure and heat of compression created a fossil fuel—natural gas—recently tapped by a human being's drill. Driven by pressure, gas came to the surface and entered a pipe, to be distributed to a gas pipeline company, the pipeline periodically cleaned with water. Eventually it got to my energy supplier, in my town, delivered to my street and my house, to heat the water in which I showered. Think about how water was involved in every step of the way, including generating electricity with steam or hydropower, and the shower! How wonderful!

Garbage Disposals

Garbage disposals can wreak havoc with sewer pipes. This is especially true for restaurants and homes that use lots of bacon and other animal fats. Those, along with other disposed wastes, form deposits in the sewer lines that are very much like cholesterol in your arteries. They get together and clog the drainage system. No, don't flush Lipitor down the drain to dissolve them! At present, municipalities and sewer district administrations must regularly clean out the pipes, a process that can cost millions of dollars. Some communities are banning installation of new garbage disposals entirely; others merely discourage their use. Dispose of your food and fat wastes via the garbage and landfills. As long as you do not put animal wastes in your compost pile, it won't smell or attract local wild and or domestic meat-eating animals. Better yet, find a new way to reclaim, re-use, or constructively dispose of them. Best is to build a simple compost heap in your yard, or use a commercially available composting device, or your local recycling agency may more efficiently and effectively handle their more obnoxious characteristics. Or simply cut down on such wastes. Remember, too, that using the garbage disposal to eliminate waste materials means that you may not be conserving natural resources in a large number of ways, and it may be most environmentally sustainable if we all are more careful about our valuable natural resources.

Waste Treatment

Water from industrial, agricultural animal or vegetation production processes may require waste treatment and a discharge permit before being released to the environment. That, of course, may be because chemicals have combined, creating undesirable, smelly, or poisonous substances used or simply dumped down the drain after using the water. Even household wastewater may contain contaminants that could make you sick, impart a bad taste, color, or odor to the water. Those impurities—or pollutants—might come from washing machines, dishwashers, toilets, or simply the house pipe system where lead in older house plumbing could be quite dangerous, especially to children and life forms in wetlands, streams, or lakes. Primary treatment involves screening or pumping to leave behind floatable and settled materials, and aeration to get rid of gases, often by making use of oxygen to combine harmlessly with the pollutants. Secondary treatment removes settled sludge and floating scum, and tertiary treatment involves killing disease causing-bacteria and removing or digesting any remaining solids. What is finally left may be environmentally acceptable by digesters, long-term filtering on land, or use as agricultural or lawn or garden fertilizer. As gray water, it may be used on some city parks, where it also helps recharge local groundwater supplies. The water might need treatment simply because it is warmer than when it came from the system supplier, stream, or well, of because it was used to cool the air in your house.

TMDLs

You'd be pretty mad if sewage came out of your kitchen water faucet; and rightly so. Most of the water pollution cleanup these days takes the form of what are called TMDLs, which stands for Total Mean Daily Load. The term refers to a pollution control strategy designed to limit the amount of a specific pollutant that enters an impaired water body. Each state, in cooperation with the federal Environmental Protection Agency, must establish and maintain the status of all water bodies in the state with regard to whether the water meets the standards for which it is used, such as drinking, fisheries, manufacturing, and so on. If they don't, they are listed as impaired, and demand special attention and action. TMDLs normally refer to end-of-a-pipe point sources of pollution. Nonpoint sources of pollution normally don't require a permit from the EPA or the state if discharge is authorized by the EPA, but now some are regulated under TMDL rules because the pollution—even though it may be diffused over the land as nonpoint sources are defined—is a potential threat to human health. Some examples are certain agricultural land uses that produce serious pollution to streams and lakes, especially when it rains or snows melt. A concentrated animal feedlot operation (CAFO) is an example. These Best Management Practices now require a permit, which means that the CAFO needs to be designed, built, supervised, and inspected by licensed and trained personnel.

Best Management Practices

I bet you've seen Best Management Practices—BMPs—as you walk or drive around your neighborhood, park, or a construction site. BMPs are used to prevent and control water pollution from nonpoint sources, land use pollution diffused over the landscape. They were first identified in the 1972 Water Pollution Control Amendments, the Clean Water Act. Section 208 required all the states, in cooperation with the Environmental Protection Agency, to create pollution control plans for urban, agriculture, forestry, mined land, waste disposal, saltwater intrusion, and construction sites. The first BMP was New York City's Pooper Scooper Law requiring pet owners to keep city sidewalks (and your shoes) clean. You've probably seen another early BMP, the beehive-like buildings that cover piles of highway salt and prevent its erosion and water pollution. A common construction area BMP is the silt fence, a semi-porous black plastic about eighteen inches high held in place by wooden stakes. Silt fences prevent sediment in runoff from a construction site going to a stream. An educational BMP is the sign next to a street drain stenciled, "Flows to Fish Creek." BMPs are important ways to contain nonpoint sources of pollution and to keep our water resources clean.

Point/Nonpoint Pollution

Sometimes simple terms are used to explain complex topics, as with point and nonpoint sources of water pollution. The terms are defined in the 1972 Water Pollution Control Amendments, the Clean Water Act. Point sources are end of the pipe discharge of municipal or industrial wastewater, and require a permit. Nonpoint sources are from land uses and are spread out or diffused over the landscape, although sometimes they may actually be collected in pipes, such as a highway's water collecting culvert. This Section of the Act is titled "Areawide Waste Treatment Management," but it is known simply as Section 208. It identifies seven nonpoint sources: agriculture, silviculture (forest management activities), mining, construction, salt water intrusion, residual waste, and on-land disposal of wastes, or landfills. By the 1980s, nearly all the nation's point sources were cleaned up, and most pollution came from nonpoint sources, with agriculture and urban runoff heading the list. The Soil Conservation Service (now the Natural Resources Conservation Service) worked with the EPA, state, and local agencies to identify and eventually require consideration of Best Management Practices, very often easy timing or policy changes, or simple structures to control pollution from barnyards, roads, cultivation, and construction. Each state has a manual of BMPs, specifications about their construction, deployment, and cleanup expectations.

Permits

Local, state and federal agencies use water quality control permits to stop water pollution. Permits were first used in the 1899 federal Rivers and Harbors Act that controlled dumping hard fill in navigable streams, requiring a Corps of Engineers' permit. The 1972 Water Pollution Control Amendments required permits for point sources of pollution, where a municipal or industrial pipe discharges wastes into a water body. That act also controls filling in or modifying wetlands, requiring a permit when a court decision declared wetlands tributary to navigable streams, thus regulating them. Nonpoint sources of pollution—polluted waters from land uses—did not require permits. Times change: permits—usually granted by local or state authorities—were declared necessary for suburban development because it reduces the land's ability to absorb rain and snowmelt and therefore contributes to polluted stormwater runoff. Stormwater permit standards may be regulated by local, state, and national agencies, but local government agencies issue the permits. With improved understanding about water pollution, more and more nonpoint sources have become point sources. Thus permits are now required for land uses formerly considered nonpoint sources of pollution. A 1994 court decision that defined a manure spreader as a point source meant that permits are required for discharges from large animal farms, such as for spreading stored manure as fertilizer on pasture lands. Many other land uses require permits because they affect total mean daily load of pollutants, or TMDLs.

Concentrated Animal Feedlot Operations

Big farming operations feeding large numbers of livestock before shipment to markets may be referred to as Concentrated Animal Feedlot Operations, or CAFOs. Prior to 1994 such establishments were considered as nonpoint sources of pollution, not subject to state and Environmental Protection Agency permit processes under the 1972 Water Pollution Control Act, or Clean Water Act. Because excrement from large numbers of animals is usually spread on farmers' fields as fertilizer by a manure spreader, preferably in the spring when nitrogen and phosphorous are directly used for growing plants, that equipment was the focus of a lawsuit. In 1994, the Supreme Court refused to hear a lower court decision involving Southview Farm in western New York that considered the typical manure spreader to be a point source of pollution. The decision meant that a CAFO needed a permit for the manure spreader, no longer a nonpoint source of pollution. The Natural Resources Conservation Service immediately went to work defining how many animals constituted a CAFO and the nature of the measures that must be taken to have the water quality pollutants in the manure neutralized, diluted, or removed from the animal wastes before being spread on the fields. The NRCS also established the process of designing and constructing the CAFO facilities to ensure continued clean water as well as the criteria for certification of the professionals who design and certify the many types of structures and plans.

Mining

We don't normally think of water use in mining—extracting minerals from the earth—but water plays vital roles in many aspects of mineral availability. Panning for gold comes to mind, of course, where water is used to float or wash away the lighter particles suspended in streams or miner's pans, leaving the heavy gold behind. Sometimes, gold was separated from unwanted rock by pumping large quantities of water at high pressure to accomplish the removal of the heavier particles. Those needs for water also became the basis for western water law, which specifies that the right to use water requires taking it out of a stream and using it beneficially and continuously. The open pit mines caused a lot more environmental damage to the landscape, damage that may still be seen more than a hundred years after the mining took place; they often produce excess runoff during rain or snowmelt, resulting in man-made flooding and pollution. Management of water in mines is also a challenge. Mine flooding is a dangerous hazard, of course, and accumulated water has to be pumped out for miners' safety. Again, western states water law recognizes that the pumper "made" the water that is now a stream, and thus has the first right to its use. Water is often used in the process of refining or cooling the mineral being mined and, finally, various mineral-bearing ores are shipped by barge on major waterways.

Navigation

Many lasting local canal systems—our early highways—were built in the days before our current highways. The 360-mile Erie Canal was the engineering marvel of the early nineteenth century. Navigation became one of three major missions of the Corps of Engineers. The others are flood control and environmental restoration. The 1824 General Survey Act enabled roads and canals; flood control was added in 1936, and environmental restoration was in the 1990 Water Resources Development Act. All three are intimately connected. Navigation is for defense, pleasure, or commercial purposes. Flood control requires dredging sediments and building dams. Dams require construction of locks allowing traffic that changes natural flow patterns and river characteristics. Soon, natural levees and islands disappeared as the rivers' flows steadied, the goal of dam and lock construction. The Mississippi, Missouri, Ohio, Tennessee, Columbia, San Joaquin, and tributaries make commercial seaports out of landlocked cities, enabling movement of raw and manufactured goods. As current six-hundred-foot-long locks are upgraded to twelve hundred feet, permitting longer barge tows and shorter shipping time, the Corps focuses on restoration of the floodplains, islands, and wetlands, characteristics of the major waterways prior to the development of the steamboat two hundred years ago.

Fighting Fire

Water, of course, puts out fires, but water plays another important role in fire control. When an insurance company makes fire insurance available to homeowners and businesses, it quotes a rate for the annual premium based on the community's fire rating, under the Public Protection Classification Program administered by the International Organization for Standardization or ISO. Premium pricing is based on an appraisal of the municipality's capability to rapidly put out a fire. Twenty-four percent of that evaluation is based on the personnel of the fire department, including training. Twenty-six percent is based on fire fighting equipment. And, since the efforts of the fire fighters would not be effective without a well-distributed and plentiful water supply that is 40 percent of the rating system. A modern tanker truck's hose can deliver up to two thousand gallons of water per minute, so the supply has to be considerable. Communities have ratings of from one to ten: one is the rate with the lowest fire insurance premium. Only about fifty-seven U.S. cities have number one ratings. San Francisco was the first. It has a completely separate salt water supply system from San Francisco Bay. The pipes are only in the downtown area and are separate from the regular water supply pipes, because salt water is corrosive. The system can be isolated from the regular water supply. What is the rating for your home or business?

Dams

Humans have built dams for hundreds of years; about 75,000 worldwide. Nearly two-thirds are high dams, over five meters. Dams are not new: beavers—and perhaps other animals—have been building them a lot longer than we have! There are several differences, however. Beaver usually build dams in open and flat stream reaches. They are low—under five feet or so—and porous: they allow the water to flow through them. Also, beaver dams provide safe access to and protection for their home, and provide means for transporting the food and construction materials the beaver chew down. Our dams are higher, impervious, and for purposes such as navigation, flood control or river regulation, domestic or commercial water or canal supply: turning a water wheel for grinding wheat in a mill or for electricity generation. High dams interfere with fish migration and downstream channel characteristics, usually adversely affecting the river's contribution to biodiversity. Water at some elevation has potential energy and it must either expend that energy by moving sediment or turn it into heat energy, sometimes both. Thus, a dam interferes with the river's natural ecology and Earth's water cycle; if the river returns to its old channel, or the dam washes out, highways and bridges, crops, buildings, and lives are destroyed. There have been some pretty spectacular and destructive dam failures. In 1972, Congress assigned the safety responsibility to the Corps of Engineers after the spectacular Baldwin Hills Dam failure in southern California.*

*Upon hearing the broadcast on 2/22/09, I noted a slight duplication in the form of "hydropower" and "electricity generation." I changed the written version to just have one, and added "canal supply," which I had not included (although "river regulation" is included, too).

Policy on Dams

Civilian policy and management are what strategy and tactics are to the military. Water policy is an important instrument in our survival toolkit. Built to provide flood control, irrigation, and municipal and industrial water supplies, dams are a classic example. The 1936 report that justifies federal flood control dam-building policy cites another government document: "The ideal river, which does not exist in nature, would have a uniform flow."* Having a constant and reliable supply of water sounds good, but my view is that dams are assaults on nature; natural rivers are essential to our sustained existence in the natural environment. Rivers are analogous to the flow of water through a body; blocking it would destroy the aquatic flow-through system. Stopping a river's natural flow is bad policy, and badly justified dam building is bad management. Rivers play an essential role in recycling of energy, soil and sediment, floodplains, waste products, and life forms. We certainly can't live without important services to our life styles that some dams provide, but the impact of thousands of dams on the natural, irregular delivery of water on the watershed and the stream systems is not contributing to sustainability. Recent examples of dam removal and artificial flood releases show the rapid return of natural ecological conditions. What do you think should be our policy on dams?

---

*C. S. Jarvis et al., 1936. *Floods in the United States: Magnitude and Frequency*. Water Supply Paper No. 771, Geological Survey, USDI, Government Printing Office, Washington, DC. Interestingly, the quote is a partial one from a statement written by Gilbert White (personal verbal communication at lunch, July 1986), Chairman of the Mississippi River Basin Commission that studied the 1927 floods and what the federal government might do to constrain their undesirable effects.

Removing Dams

Flying in a light plane over California's north coast, I noticed where a stream that normally discharged into the ocean was blocked by a sand and gravel bar created by the relentless California Current flowing southward. I wondered how the salmon swam between the stream and the ocean and realized that the definition of the mean annual flood was *bank full capacity,* a flow that is topped every two to three years. Thus, about every two to three years a streamflow is big enough to breach the barrier bar, allowing the fish to chemically find the stream of their birth and return to it for spawning. That probably is why salmon, shad, and other similar species have two to three-year life cycles, and why so many of these fish that spend some of their lives in both ocean and stream have disappeared: our flood control dams prevent that natural breaching of the gravel bars. Recent removal of dams built long ago for transporting logs from tree harvesting operations, flood control, water supply, and hydropower, has resulted in astonishingly fast restoration of fisheries in the northeastern states. People had hoped that the first dam removals or change in operation procedures would result in fisheries restoration in a couple of years or so. The fish have actually returned in a matter of months. Now we have to decide which dams to remove, or have useable fish channels built around them.

Water Project Economics

Water projects are expensive. We need to build high-cost flood control, navigation, and water supply systems and pay for them later. The first or fixed costs of levees, canals, pipelines, water treatment plants are high, so we get cash now. We facilitate government or private utilities borrowing money from the future so that we can have abundant, safe, and low-cost water today. This is accomplished by issuing bonds. This sounds like a definition of conservation: shifting rates of use toward the future. But it isn't exactly the same, and there are three unintended consequences: First, water is undervalued: we don't pay its full value. Second, we burden our descendants with the costs of providing us with water now. Part of the problem with water and energy is that both are subsidized; in other words, we don't pay the full cost. Our kids have no choice about it. This applies to other bonded projects too: environmental protection, roads, bridges, airports, and energy. Third, since everyone in any local service area pays similar water rates, the poor pay a greater percentage of their income for water. That is a social justice issue. A simple way to correct these problems is to raise the price. But what sounds good, is not good for cheap, abundant, and clean water now, and, again, the poor pay disproportionately. These are important issues that we need to talk about to arrive at acceptable, equitable, and effective solutions.

Conservation

Water Conservation is important to *everyone*. It means using water wisely, not just saving it. The term *conservation* was first applied to natural resources by Gifford Pinchot, advisor to Theodore Roosevelt, who appointed him first chief of the U.S. Forest Service in 1903.* Pinchot wrote that forest conservation was "the greatest good for the greatest number in the longest run." Franklin Roosevelt appointed Hugh Hammond Bennett first chief of the Soil Conservation Service in 1933. Bennett's practical conservation focused on soil erosion control and erosion prevention during the Dust Bowl of the 1930s. Soil loss severely damages air and water quality, flood behavior, and agriculture. In 1955 another conservationist, Siegfried von Ciriacy-Wantrup, an agricultural economist, defined conservation as "shifting rates of use towards the future." To me, conservation is a spectrum of management policies and practices with exploitation at one end, all use and no time, and preservation at the other end, all time and no use. Since our water resources serve all society, we all need to be involved in conservation. Water conservation, of course, affects all the natural resources. And remember: we have the same amount of water available as when the dinosaurs roamed the Earth.

*The Forest Service was created in 1905, but Pinchot was appointed head of its predecessor(s) the Division and later the Bureau of Forestry and the year is uncertain.

Center Pivot Irrigation

If you look out of an airplane window as you fly across the western plains states, you will see round green fields. No, they are not space invaders, these are high tech, efficient center pivot irrigation systems, shaped by a surface or groundwater supply distributing water through pipe mounted on fifteen-foot-high wheeled A-frames. A computer controls system rotation timing, including wheel speed, the rate of water delivery, and whether and how much fertilizer, insecticide, fungicide, or weed killer is applied. The system operates at night when evaporation loss is lowest and the greatest amount of water gets to the dry soil. The fields are often one-quarter of a square mile, or one hundred sixty acres, thus the pipe has to be about a quarter mile, or 1,320 feet long, about thirteen one-hundred-foot lengths of pipe between the A-frames. Sometimes you will see only half or a quarter of a circle irrigated, or a wedge-shaped piece left out because that is where a homestead is located. The corners of the fields may be irrigated by elbow extensions of the pipes, or those may be left as important wildlife areas. The center pivot irrigation system allows sloping land to be irrigated, which can't be done with regular flooding irrigation, and the high pipes can easily pass over the highest cornstalks and even oil pump installations. The water can be delivered in a spray or drip system, which saves water.

Government, Law, and Organizations

Organizations

It can be frustrating trying to reach the right government agency to solve a water problem. If you add in private organizations, the confusion is amplified. Thousands of such organizations were created, mostly during the twentieth century, and exhibit varying formality and authority, purposes, and serve different interests and groups. Government organizations include land and natural resources management, regulatory, construction, enforcement, research, and study purposes. Some have especially important responsibilities for water, such as public health agencies. The states manage state-owned natural resources, and often coordinate with more broadly involved federal agencies. Regional organizations such as river basin commissions may cover several states and may involve Canada or Mexico. Professional organizations provide information, standards, and licensing or certification. NGOs—nongovernmental organizations—provide opportunities for citizens to interact and to play an essential role in the policy and management decisions for the vital water resources we all use. When different organizations and individuals get together, they form partnerships, the most important of all of our thousands of water resources organizations. If you wish to find an organization to answer a specific question, it might be best to start with a call to your local soil and water conservation district.

Federal Organizations

Finding out what government agency is responsible for some complex water resources policy or management issue in which you are vitally interested can be exasperating. Let's sort it out. Federal agencies with water resource responsibilities are of several types: some manage lands, such as the Bureau of Land Management, Forest Service, Park Service, and the Fish and Wildlife Service. Others are primarily construction agencies, including the Corps of Engineers, Bureau of Reclamation, Natural Resources Conservation Service, and the Tennessee Valley Authority. Another group includes regulatory and enforcement agencies: the Interstate Commerce Commission, Federal Power Commission, Flood Insurance Administration, and the Environmental Protection Agency. Research and Inventory agencies include the Geological Survey, the National Weather Service, Agricultural Research Service, and the National Oceanic and Atmospheric Administration. Lastly, there are often short-lived study and coordination organizations in the administrative or legislative branches, the reports of which are valuable guides to future water resources management, such as the still-pertinent National Water Commission's 1973 informative report. In New York there are also the state departments of Health, Environmental Conservation, Agriculture and Markets, Parks and Recreation, and State, which is responsible for administration of the Coastal Zone Management Act. A most interesting and effective organization in New York is the State Soil and Water Conservation Committee. From its home base in Ag and Markets, it coordinates soil and water conservation policy in the state, including coordination of the fifty-eight county soil and water conservation districts.

Land Managing Agencies

You may have visited national park or forest lands. Several federal agencies manage those lands for renewable resources. The federal government started forest land management at the end of the nineteenth century when forest lands had been cut over and abandoned as the timber companies logged and moved west. Disastrous floods followed, especially in Pittsburgh and Johnstown, Pennsylvania. Research by the Forest Service and Weather Bureau at Wagon Wheel Gap in southern Colorado scientifically linked timber harvest and excess runoff providing justification for the Forest Service. Prior to that, early forest reserves and parks had been set aside, without management. Under the jurisdiction of the National Park Service and the Bureau of Land Management, public domain lands left over from the Louisiana Purchase, wars, and land treaties for which the land was too steep, dry, or shallow-soiled to farm, became parks, monuments, or grasslands. The Bureau of Indian Affairs assists Native Americans in management of renewable resources on federal reservations west of the Mississippi River, and has had changing policies over the years. Many Bureau of Land Management lands provide minerals and are too dry for any crops. Agencies grew, changed names, moved from one federal department to another, and learned more about vegetation, land, and water management. While small in total area, the National Parks—along with the National Forests—are particularly important because at high elevations they provide lots of annual runoff—mostly snowmelt water to our major rivers.

Politics

Although the professional natural resource managers of our federally controlled lands operate in the long-term interest of the American people, the leaders of the agencies—Agriculture, Land Management, National Forests, National Parks, Wildlife Refuges, and so on—are political appointees and, if not also career professionals, are often under pressure from presidential leadership or policies by congressional budget control processes. The result may be termination or radical changes in longstanding protection, redefinitions of sustainable management, or outright termination of land status that protects water supplies, endangered species, scenic wonders, and most important, protection and enhancement of our wildland ecology, the biodiversity that is essential to our collective future and well-being. Biodiversity is the summation of the wealth of plants and animals that provide the resilience we humans need to ward off disease, as well as variations in climate such as drought, flood, wildfires, volcanic eruptions, invasive species, erosion, and pollution. Even with a seemingly long four-year political occupation of the White House, lasting damage to long-term management of slow-growing vegetation and slow response to hastily made management decisions can wreak havoc with our futures on the planet. Examples include termination of long-term preservation status or on the ground changes, altering regulations about resource extraction such as minerals and fossil fuel mining, and long distance water diversions that impact either or both the source of the water supply and the place where it is used or discharged.

Bureau of Land Management

The Bureau of Land Management administers more than 50 percent of the United States' public lands, about 470 million acres, for you and me. Those are lands that the nation acquired mainly from France, Russia, and Mexico by purchase or conquest. Generally, they do not receive enough water to grow crops, are too steep, or are not valuable for much except mineral resources; they were left over from the General Land Office, created in1803 to administer the Louisiana Purchase, and the Taylor Grazing Service, created in 1934 to regulate grazing leases on public domain lands. The two agencies were combined in 1946 becoming the BLM. Its lands are administered under a 1964 Classification and Multiple Use Act, similar to that which directs the Forest Service. Many BLM lands are the wide open spaces of the western United States west of the Mississippi River, so its headquarters are in Denver, not Washington, D.C. Some of the lands are grazed and others are mined. Most are arid, but there are some water resources. Often federal lands are traded between agencies to provide better management, especially where water is concerned. The lands include about 2.6 million acres of excellent Douglas Fir forest in western Oregon, so the BLM has major forest management and wildfire control responsibilities there, and 98 percent of Alaska is under BLM jurisdiction, including many valuable water resources and associated forests, lakes, rivers, and glaciers.

Forest Service

The U.S. Forest Service was created in 1905 from the Bureau of Forestry. It is responsible for the management of all renewable resources on nearly two hundred million acres. Its first Chief Forester was Gifford Pinchot, an advisor to and sparring partner of President Theodore Roosevelt. Initially, the Service was in the Department of The Interior, but was transferred to Agriculture to manage the Forest Reserves set aside since 1891. Major flooding followed severe and unplanned timber cutting in the nineteenth century. Those lands were abandoned and then made part of the National Forest system through the 1911 Weeks Forest Purchase Act. Thus, the federal government got into forest management competing with and then challenged by the forest industry. The Weeks Forest Purchase Act was made constitutional by the opening phrase: "In order to protect the flow of navigable streams" and, since courts have defined "navigable" as being capable of floating a canoe, virtually all waterways are considered navigable. That ensured that the forested watersheds, presumably of any stream, came under the Weeks Act, as well as control over navigation. Pinchot's ideal was to manage the forests "for the greatest good for the greatest number in the longest run." To do that, the Service has four branches: Administration, Research, State and Private Forestry, and International Forestry. It cooperates with state and local fire control units, conducts forest research all around the country, and has been involved in numerous controversies over the amount and style of timber cutting, who harvests the timber and at what price, and grazing and recreational uses. The high elevation forests mean that the Forest Service supplies good water too.

National Park Service

We have some wonderful treasures in the National Park System. The agency that manages natural and cultural resources of the nation's parks, monuments, recreation areas, and historic sites is The National Park Service. It was created by Congress in 1916 after arguments between interests led by Gifford Pinchot, who favored use over preservation of many natural resources, and John Muir, who favored preservation over use. The congressional action followed the 1906 San Francisco earthquake and fire, which clearly indicated that the city needed the Hetch Hetchy dam and reservoir, a new city water system. The city also preferred that the public believe the disastrous fire was avoidable by having an adequate water supply, not by an earthquake that no one could control. The Park Service manages for preservation, research, and access for public enjoyment of our treasured sites, including unique natural or historic areas such as the Statue of Liberty, National Mall, notable landmarks and battlefields, Mount Rushmore, and Mammoth Cave. Many parks are located adjacent to National Forests and other federal lands, and conflicts and better management opportunities often arise, resulting in land exchanges for more effective management. Of course, I wouldn't be talking about National Parks unless they were of particular interest to our water resources! Many of them are high-elevation lands that, although they are small in total area, receive lots of snowfall, and thus nourish many of the nation's rivers. And some, such as the Grand Canyon, were *created* by water.

Fish and Wildlife Service

Before doing what you want with public waters, one agency you will have to deal with is the U.S. Fish and Wildlife Service. It was created in the Department of the Interior in 1940 from the 1885 Bureau of Biological Survey and the 1939 Bureau of Fisheries. The Fish and Wildlife Service's duties expanded as we understood the importance of fish and wildlife to our sustainability. The agency administers about 10 percent of our federal lands, 94 million acres of wetlands and estuaries, essential parts of principal north-south flyways of migrating birds, and integral parts of regional water resource systems. In 1946, the Fish and Wildlife Coordination Act required anyone to inform the Service that they planned construction in or near federal lands and waterways thereby providing the Service with the opportunity to respond to the potential threat to fish and wildlife of draining land, dumping, dredging, damming, straightening channels, or diverting water to or from a water body. As with most federal water resources legislation, these requirements are restricted to navigable waters, ensuring constitutionality. The courts generally consider that all tributaries to navigable streams are also navigable, including wetlands. Also, since many fish and wildlife species are migratory, they cross state boundaries, and are therefore under federal jurisdiction. The Coordination Act was a major step enroute to requiring federal agencies—and state and private entities—to prepare environmental impact statements based on interdisciplinary surveys of potential environmental changes by proposed construction.

Corps of Engineers

The Army Corps of Engineers is the oldest federal water resources management agency. Its civil works program also has the broadest set of responsibilities, involving navigation and defense, floodplain management; flood control mapping, construction, and emergency cleanup; water supply and recreation; dam safety; permits for discharge into navigable streams and their tributaries; water supply, and environmental restoration. Congress passed laws assigning these activities from 1824 to 1992. Until about Earth Day in 1970, the Corps was more concerned about water resource management engineering solutions than environmental issues. Its three thousand personnel and multiple programs have become more environmentally responsible as it pioneered establishment of broad partnerships for water resources management. Since the late 1960s, Corps management and construction plans have been reviewed by its Environmental Advisory Board, on which I had the pleasure and honor of serving for four years. Environmental concerns often become secondary to flood damage control, rescue, and repair, producing antagonism to Corps activities. The Corps constructed and operates many large flood control projects and associated navigation locks, and it manages water resources on military reservations around the world. It is engaged in toxic substances and brownfields cleanup in cooperation with the Environmental Protection Agency, and also contracts for work with foreign governments when requested.

Bureau of Reclamation

If there were an agency with "reclamation" in its name one would think that it would do some sort of restoration. There is such an agency, and it doesn't do restoration. The federal Bureau of Reclamation was created in the Department of the Interior by the Reclamation Act in 1902. It gave the BR responsibility of providing irrigation water to dry western public domain states. Major portions of the west are quite arid, and bringing water to them doesn't restore or reclaim the water: it just moves it from here to there. However, the BR has provided the opportunity for lots of agricultural production, jobs, an improved standard of living, and has contributed much to the national economy. As the BR builds dams, tunnels, and canals to bring the water to dry lands—even across the Continental Divide—water is transported from one watershed to another for crop irrigation; most of it is evaporated and transpired by the vegetation, and irrigation is our largest consumptive water use. It reduces the source river's flow, its flood peaks, and alters major ecological functions of stream waters. While the BR has enriched the nation's economy, it has simultaneously spoiled some very important aquatic environments. The BR has also established productive working partnerships with local districts to manage funds and water rights to improve local agriculture. Currently, work is underway to mitigate the adverse effects of many smaller dams by modifying them or their operation procedure, or even removing them.

Natural Resources Conservation Service

Created in 1994, the Natural Resources Conservation Service—NRCS—was first known as the Soil Conservation Service, created in1935 from the Soil Erosion Service, which was known before 1933 as the Bureau of Soils. It has been in the Department of Agriculture except when it was briefly transferred to the Department of the Interior to manage submarginal public domain utilization projects. Originally a service-and-education agency, the NRCS provided soil and water management advice to farmers; about how, what, and when crops ought to be planted, managed, and harvested. The 1985 Farm Bill required that the NRCS certify if new wetlands or highly erodible lands were brought into production, making NRCS more regulatory. It also had construction responsibility following the 1936 Omnibus Flood Control Act, which assigned upstream flood prevention to the SCS and downstream flood control to the Corps of Engineers. The NRCS is structured by state, with a state conservationist and employees—soil scientists, engineers, hydrologists, foresters, managers, economists, and other specialties—located in several federal, state, and local fiscal and management agencies with the soil and water districts. In some states, the districts are organized on a watershed basis; in others they are defined by county boundaries. NRCS welcomes public participation, so stop in.

Tennessee Valley Authority

The Tennessee Valley Authority (TVA) is one of four major federal construction agencies. It differs from the Corps of Engineers, Natural Resources Conservation Service, and Bureau of Reclamation in that it covers a specific geographic area, the watershed of the Tennessee River, tributary to the Ohio. The TVA was established in 1933 as a part of President Franklin Roosevelt's programs of getting people back to work in the Great Depression, through conservation. Congress created the TVA to bring a severely underdeveloped portion of the Nation to a higher economic level, making it productive, providing roads, education, and energy, also necessary for atomic bomb development. That notable effort is reported in a book by the TVA's first director, David E. Lilienthal, entitled *TVA: Democracy on the March*. Congress justified the mammoth expenditure to build more than thirty flood control, power, and navigation dams, roads, manufacturing plants, and schools in the name of national security because the Authority provided both sites and power for the Manhattan Project, the code name for the Atomic Bomb development. Nevertheless, the TVA did socially and economically what it set out to do by harnessing the hydroelectric power of the valley, providing 650 miles of navigable channel, and helping control serious flooding on the Tennessee, Ohio, and Mississippi Rivers. The TVA also conducts regional watershed research and, unfortunately, has been the nation's biggest consumer for environmentally damaging strip mined coal, also used for electricity generation.

Environmental Protection Agency

The Environmental Protection Agency, or EPA, was created by President Richard Nixon's Executive Order that combined parts of several other agencies, in 1970, the year of the first Earth Day. It is an independent agency, in the office of the president, and has five branches: Air and Radiation, Toxics, Solid Waste, Water, and Research and Development. It has major responsibilities that relate to water in both natural and polluted settings. The EPA is responsible for the administration of the Water Pollution Control Amendments, amended thrice since the landmark amendments in 1972 to the 1949 Water Quality Act, now known as the Clean Water Act. It has turned over some of the process to state departments of conservation, environmental quality, or similar names, and shares the administration of wetland permits with the Corps of Engineers, the Natural Resources Conservation Service, and the Fish and Wildlife Service. The EPA has responsibility for cleaning up brownfields, the sites left behind when a polluting industry closes and leaves its waste behind. Because the EPA also includes the former Public Health Service, it is also responsible for water quality standards, working with local units of government, especially state and county health departments. The EPA also monitors the environmental impact statement process, and works closely with the Council on Environmental Quality and the U.S. Corps of Engineers. As a regulatory agency, the EPA can prosecute violators or water quality standards and levy fines.

Geological Survey

Boaters, anglers, anyone who uses a river for sporting activity owes a tip of the hat to the U.S. Geological Survey. The Geological Survey is one of several federal agencies that conduct research, and it maintains historical and real time records of streamflow, such as surfing conditions in the Columbia River Gorge. There have been many changes in streamflow and groundwater measurement technology in the past 120 years. The Survey has been at the forefront in setting standards for water measurement in cooperation with other governmental and educational units. It was created in the Department of The Interior in 1879 on the recommendation of John Wesley Powell, a one-armed veteran of Shiloh. He was first to navigate the Grand Canyon, and was the GS chief for thirteen years. The GS is responsible for mapping the Earth, planets and their moons, for which it cooperates with the National Oceanic and Atmospheric Administration. It produces maps of land, geologic, and mineral resource surveys and has been particularly involved in water since 1894 when it started the nation's stream-gauging program. It conducts water quantity and quality research on both surface and groundwater resources. It works particularly closely with about twenty-five additional federal agencies and with state water resources agencies. Many of its offices are located at colleges and universities. Most recently, the Survey has established benchmark watersheds around the country that provide important baseline data for detecting changes to our water resources.

NOAA (and NWS)

The two great buffers of planet Earth are the oceans and the atmosphere. The National Oceanic and Atmospheric Administration—NOAA—created in 1970, combined responsibilities of several already existing agencies, but—at the start of the Environmental Decade—with a new and better formulated scope. It now includes the National Weather Service and the National Marine Fisheries Service. The largely freshwater fisheries are under the jurisdiction of the Fish and Wildlife Service. "NOAA conducts an integrated program of management, research, and services related to the production and rational use of living marine resources, and protects marine animals."* NOAA is also responsible for control of ocean dumping and marine and energy resources, and administers the research-supporting national Sea Grant Program. The National Weather Service used to be called the Weather Bureau and was created in 1890 in the Army Signal Corps. From 1965 to 1970 it was combined with the Coast and Geodetic Survey in the Environmental Services Administration. Obviously, if you want to look up weather data, you need to know when the data were collected so as to look up the right agency. The NWS is responsible for meteorological data collection, analysis and research, and for real-time flood forecasting. Individual station data are published in the national daily weather map, as electronic data banks, and for remote data transmission. Monthly data are published in the publication *Climatological Data*. Weather data, of course, are essential inputs to water and watershed management.

---

*Federal Register (980)

Districts

Did you ever notice the state and county boundary lines on a United States map? Eastern boundaries are more squiggly, set by surveyors on natural boundaries like streams or ridges. Western boundaries are more often straight. The Natural Resources Conservation Service works closely with local soil and water conservation districts through a federally employed State Conservationist. It is gradually shifting its operations closer to the districts than the state. That means the NRCS will be more district oriented. This change opens the possibility of a better working relationship at the local level, which might actually be better for future soil and water conservation and resource management, simply because the district is closer to the natural resources and to the people who live on the land. In the western states, district boundaries are closer to local, regional, or multistate watersheds, although state boundaries are not. Thus, overlap of natural-based governmental district units and political boundaries of states and counties often make it difficult to make better and more effective conservation and management. The reasons behind the shift in federal agency operations are fiscal, notably cutting costs and streamlining the agency's—and the districts'—operations. I could be all wet on this one, but change is inevitable, irresistible, and maybe for the overall good of conservation. As one of the New York State Soil and Water Conservation Committee members said, "Don't shoot the messenger."

Interstate Water Law

States fight over water rights just as people do. The first interstate water case was *Wyoming v. Colorado,* over the Laramie River, which rises in Colorado before flowing into Wyoming. In 1922 it established that where each state internally recognizes the appropriation doctrine—that the first water user in time has the first right—could apply between states, too. This decision enabled the Colorado River Compact, which has no watery connection with the Laramie River: but the precedent did. Six of the seven basin states approved the compact four days later. Arizona didn't like the compact's water allocation and immediately sued California. Arizona lost three times before succeeding: four Supreme Court cases lasting more than forty years, with the amount of Native American Water Rights issues extending into the 1980s. A related case was *Nebraska v. Wyoming,* settled in 1945. Its importance lay in a claim by the federal government that set the stage for the reservation doctrine, which maintains that the United States has rights to water in public domain states if it didn't dispose of them when granting statehood, and that public land uses have water rights. Thus, trees on National Parks and Forests lands have reserved water rights. Colorado didn't like that idea, since Congress had approved its constitution in 1876, bequeathing its water to the people. Colorado failed. Even today, reserved water rights for many public domain federal reservations are being quantified.

Native American Water Rights

Native American water rights have been an important issue for more than a century. When members of the Blackfeet Tribe were confined to the Belknap Indian Reservation on the Milk River in Montana Territory in 1888, they diverted river water for irrigation. Later, upstream settlers diverted river water under the Appropriation Doctrine, which asserts first in time is first in right. The tribe sought support from their Indian Agent to recover the reduced flow. Risking his life, the agent agreed that the tribe should have water, and secured agreement from a U.S. Marshal and a judge who, according to David Getches, "did what they had to do because it was right." The decision was appealed by upstream appropriator Winters. In 1908 the Supreme Court let the lower court's decision stand, establishing "Winter's Rights" holding that reserving land implied reserving water as well. The federal government reserved lots of land, so the Reservation Doctrine applies to all federal reservations: forests, parks, and monuments. Yes, trees have standing in court for water. The date of the reservation determines date of the water right; if the Indians were confined to their ancestral hunting grounds, the date is from time immemorial. The Winters Doctrine is federal law, whereas the Appropriation Doctrine is state law. Both apply primarily to the western—public domain—states. The issue continues to this day, since the actual amount of water for Native Americans has not yet been completely settled.

Advisory Committees

More than a century ago, the federal government created the Forest Service, requiring advisory committees for each National Forest. Living in Northern California in 1961, I served on the Six Rivers National Forest Advisory Committee, along with an ad hoc water committee for a Humboldt County supervisor. I have served on advisory committees ever since. Currently, I serve on local, state, and national advisory committees, representing my college president who can't attend them all. Advisory committees are fascinating. Members spend a day each month, quarter, or year discussing issues and specific practices, rules, laws, and new scientific information about water, leading often to new rules, legislation, or just plain good ideas about practices. It is important and active water conservation. The meetings are always interesting and I learn a lot from them and enjoy knowing the dedicated people who regularly—and sometimes intermittently—attend. Many such meetings may be open to the public or specialized interest groups that might communicate effectively with decision makers. If you have a particular interest in water and related land resources, it may already have attracted you to an advisory committee that you could attend intermittently or regularly. This is a great way to get involved and help in the important decisions about water conservation. And you can make a difference, too. It is also a wonderful way to learn about the practical challenges of water management, and how they might be managed.

Partnerships

Partnerships play an increasing role in the management of our watershed resources. These informal organizations are constructed so that all interests can be fairly and effectively represented, and all participants can feel they have had a chance to make difference in how their water and related land resources are managed. All stakeholders must have the opportunity to participate in the decision-making process; and must feel their viewpoints are listened to and valued. No one participant can exercise fiscal, organizational, political, or social control, and the leader of the partnership must be disinterested, that is, have no personal vested interest in the process or the outcome. Interestingly, the word *stakeholder* now means exactly backward of the original term: it meant a poker game participant who had withdrawn from and had no vested interest in the outcome of the current hand, and thus could be trusted to hold the stake or pot. The partnership must consider common goals and objectives that can be accomplished together better than separately and, when one set of objectives has been achieved, can agree on the next set. The first such successful partnership created by private citizens working with local, state, and national agencies was—and remains—on the Potomac River. Hundreds of others have followed, each especially adapted to the local needs and conditions. Today, many partnerships are said to really be problemsheds; and rightly so. They can be particularly flexible, efficient, and effective where formal organizations cannot.

Professional Organizations

Many people are doing a lot of work to preserve and protect our water resources. Many organizations help water professionals and the public do a better job managing the resource and serving the public. In my field, and in alphabetical order, there's the American Fisheries Society; American Geophysical Union; American Institute of Hydrology; American Rivers, American Society of Civil Engineers; American Water Resources Association; American Water Works Association; and the Freshwater Society; National Ground Water Society; and the Soil and Water Conservation Society. Usually, the word *society* or *union* means some form of certification or licensing is required for membership and/or the professional; they provide formal educational programs, workshops, and standards for certification or licensing often required by law. Associations are more loosely defined and often welcome lay public members. Most publish professional journals, and some publish popular periodicals valuable to consultants, practitioners, and the public. Almost all have annual or seasonal meetings with proceedings, educational sessions, and workshops. You can usually find these groups by their name or initials followed by *dot org*. Many of them have links to the others, and their Web sites contain information about water resources for the lay public; so do local water supply utilities and state and federal agencies.

Water Law

Laws that control how United States waters are used are varied, and reflect the region's ecology and history. East of the Mississippi River—in the colonies—plentiful water was available, and the English riparian doctrine was adopted. Land adjacent to water is called riparian, and on it, water needed for purposes such as household or domestic use, manufacturing, power production, watering livestock, and so on could be easily obtained. The region is well blessed with precipitation, so crop irrigation is unnecessary. West of the Mississippi River, however, the plains that were desirable for growing crops or watering cattle don't have much rain or snow, and the water must be transported to where it is needed. That required taking—appropriating—the water out of the stream, building a flume or canal, and using water beneficially for household, irrigation, manufacturing, or livestock watering. The precedent-setting case in the gold fields of California in 1855 was adopted into Colorado's Constitution in 1876, thus is sometimes called the Colorado Doctrine. This Appropriation Doctrine requires a genuine diversion from the stream, continuous use as well as flume or canal building, and beneficial use of the water. Those requirements don't include water for navigation, fishing, recreation, or any instream use. Recently, western law has been modified to include these instream uses, and to ensure water for Native Americans on otherwise dry reservations.

Water and Laws

Some of our most important federal water laws were enacted following some water crisis. For example, the 1978 dam safety act was passed by Congress following the 1976 catastrophic failure of the Bureau of Reclamation's new Teton Dam and the earlier collapse of the Baldwin Hills dam near Los Angeles. The original federal water quality act was approved in 1948, following early color pictures of water pollution in popular magazines of the time. Going even farther back, after the 1929 Mississippi River flood, the Soil Conservation Service was created from the former Soil Erosion Service in 1934 when a Midwest dust storm darkened the skies in Washington, D.C., prompting Hugh Hammond Bennett to remark during congressional testimony "there goes part of Oklahoma now." Still earlier, abandonment of cutover forest lands resulted in the 1907 flooding of Pittsburgh by the Monongahela and Allegheny Rivers, which prompted the Weeks Forest Purchase Act of 1911. It got the federal government into forest management, constitutionally justified because forests protect the headwaters of navigable streams. Valid federal legislation must refer to the Constitution, and since navigation is a federal responsibility, many laws involving land and water management are justified on that basis.

Watershed Initiatives

Many people feel strongly about a particular water resource and are dedicated to controlling it through a variety of watershed initiatives. Each organization focuses on particular water resource units, such as rivers or watersheds, a local problem or challenge. The resulting organization—formal or informal in structure—may be identified by a boundary, such as a watershed; a pollution cleanup challenge; natural vegetation or wildlife limits, or management concerns; fluctuating water levels; threats of urban and/or commercial development, or major construction projects, such as a highway or an airport. Ideally, the structure of the managing organization must be flexible and tailor-made for the local situation: for example, it might include one or more federal and state agencies already involved within the water challenge area, and one or more local government units—county, city, town, or hamlet—might be involved. Individual landowners (I prefer land *stewards*) usually have a financial interest in management decisions. Such stakeholders have a vested interest in the outcome of the group decisions. All stakeholders must have the opportunity to participate in the process, whether they do so or not. The initiative's leader must be disinterested in the process and the outcome of achieving the initiative's goals. If the challenge includes reaching a meeting of minds of individuals who have very different expectations of what the management outcome could or should be, a completely disinterested workshop leader needs to be employed to facilitate an outcome acceptable to all.

Five Forested Watersheds

Even with bottled water available, many folks happily drink tap water from a municipal water supply system reliably safe from disease, drought, and unpleasant odors or taste. Municipal systems provide water from lakes or rivers, desalinized ocean water, or groundwater for more than half the world's population. In the United States, there are five largely forested, surface municipal water supplies that are not filtered. That is, they cannot remove disease-causing spores such as cryptosporidium. New York City, with two thousand square miles of watersheds in the Catskill Mountains and more east of the Hudson River, is largest. Decreasing in size, the others are Boston, Seattle, Portland, Oregon, and Syracuse, New York. The middle three watersheds are mostly publicly owned and forested. The largest and smallest—New York City and Syracuse—include dairy farms on the watershed's private lands. Cryptosporidium spores are transmitted mostly by calves. Filtration is so expensive that the watersheds operate under performance criteria agreements with the Environmental Protection Agency that require active land use programs to control spores. But the cities must have filtration plant plans ready just in case the BMPs fail to meet the performance criteria. A new water treatment alternative is ultraviolet radiation that kills spores at lower cost. Until perfected, watershed management in both of New York State's unfiltered surface water supplies produces high quality, prizewinning water.

Great Lakes

The five Great Lakes contain about one-fifth of all Earth's fresh water. One half of the lakes' water is in Lake Superior. It has the world's largest area and is also the deepest, although Lake Baikal has a greater volume. Superior's six hundred feet above sea level surface drops twenty-three feet through the Sioux Locks to Lakes Michigan, Huron, and Erie. The 326-foot drop to Lake Ontario over Niagara Falls is used for power production, and is bypassed by the Welland Canal for transportation; the 243-foot drop from Lake Ontario is used for power production on the Saint Lawrence River, alongside the Eisenhower Locks at Massena. Less than 1 percent of the system's total water content is replenished each year by basin precipitation: the total lake area is nearly one-*third* of the nearly three hundred thousand square mile area of the basin. Administration of the Great Lakes basin is by the International Joint Commission or IJC, under the 1909 Boundary Waters Treaty between the United States and Canada. Thirty-three million people live in the Great Lakes Basin, and there are major basin diversions through the Chicago River for water supplies and for conveying city wastes down the Mississippi River system, and elsewhere for power production. Many communities divert water out of the lakes, but replace it under agreements with the IJC, which controls diversions out of the basin.

River Basin Commissions

River basin commissions are created to manage a river in several states and for which there may be an interstate compact or other formal agreement. Commissions, made up of representatives from the state governments administer the legal provisions. Compacts are also created for international streams, often requiring complex member relations among state, provincial, and national governments. River basin compacts—and commissions—may be for water supply, navigation, flood control, power production, and/or water quality. Some river basin commissions were created under the 1965 Water Resources Planning Act, and are called Title Two Commissions. Some, such as the Delaware River Basin Commission, focus on ensuring sufficient water based on a Supreme Court decree for the downstream states of Pennsylvania and New Jersey. The Susquehanna River Basin Commission is primarily concerned with instream water quality and sufficient deliveries to the Chesapeake Bay and its extensive estuarine resources. The Columbia River Basin Commission is concerned with managing the river for power production, flood control, fisheries, irrigation, and recreation. The St. Lawrence River Basin Commission is international, primarily concerned with power production, Great Lakes navigation, and high-quality water supplies. The most confusing interstate commission involves the Apalachicola, Chattahoochee, and Flint Rivers, in Alabama, Georgia, and Florida. The compact's complexity reflects the fact that the single river system has a different name in each of the three states.

Mississippi River

At 2,320 miles, the Mississippi River is the world's fourth longest.* That is the distance from the mouth to the most distant tributary's start. On a map, however, the named Mississippi River starts in Minnesota at Lake Itasca. Native Americans called it Messipi, meaning river. Exploring upstream from its Gulf of Mexico mouth, pioneers didn't turn left at what is now St. Louis, but went straight ahead. Otherwise, they would have gone upstream on the Missouri and Jefferson Rivers to the Mississippi's real start near the Continental Divide in Montana's Rocky Mountains. Actually, at 2,341 miles, the Missouri River is *longer*. The Mississippi watershed or drainage basin is about 1.2 million square miles. The official upper reach of the river, north of St. Louis, receives runoff from Minnesota to New York and Kentucky. By the time the river reaches the Gulf of Mexico, it carries a heavy load of silt and waste products from all of the human activities upstream. The Corps of Engineers administers twenty-nine locks for huge amounts of barge traffic. Since about 1990, the Corps has been restoring the river to pre-steamboat traffic conditions that had dramatically changed the river throughout the 1800s. This has required much wetland and river channel restoration, especially for storing and absorbing the volume and energy of floodwaters that naturally threaten New Orleans and Louisiana and neighboring lowlands.

*The river has many well-known tributaries including the Arkansas, Red and White Rivers, the Milk, Powder, Bell Fourche, Cheyenne, Platte, the Ohio, Tennessee, Allegheny, Monongahela, and all their tributaries. But that's because Hernando de Soto, the first European to document it, simply crossed it and continued westward.

Colorado River

Many rivers have great stories. I think the Colorado River's story is the best, because a surveying error made it an international stream when it barely qualifies. A 1922 court decision involving water use disputes led to the very first compact on the Colorado River. The compact between seven states, Mexico, and the United States is dated three days later. Its purpose is to divide the river's flow between growing downstream interests, while protecting future needs in upstream states, the river's source. Balancing Upper and Lower Basin strength was a challenge, accomplished by placing three states in both the upper and lower basins, and two states exclusively in each basin. The watershed boundary between the Upper and Lower Basins provided a logical dividing line. Hoover Dam was built, but only after Congress passed the 1928 Boulder Canyon Project Act that perfected the compact, which Arizona never actually signed. Other dams followed. Arizona sued California for exceeding its share, but it took forty years, the longest Supreme Court case ever. In the 1963 appeal, a Special Master's earlier notation that four Indian Reservations on the lower Colorado had water rights resulted in another twenty years of litigation. Unfortunately, the compact's annual flow of the river was determined during a wet period. Evaporation and diversions out of the watershed diminish flow further. So, the river usually runs dry before it gets to Mexico, and is too salty to use anyhow.

Closing

This is the last of about 130 broadcasts. The first appeared more than two years ago, so repeats may start, hopefully without boring you. I may have some new ones, too. This idea occurred to me in April 2005. Drafting was easy: editing drafts to 250 words for a ninety-second broadcast was a real challenge. Inspired by *Star Date* and John Weeks's musings in *The Nature of Things,* I approached WRVO with the idea, was welcomed, and am delighted to have been associated with the many Oswego and Syracuse professionals. Christopher Baycura, Producer, and David White, Media Coordinator, both at SUNY ESF provided high quality recording and editing ideas. WRVO Program Director Fred Vigeant has been a faithful contact. My wonder of water stems from learning and teaching about water, and from long-ago discussions with physicians about similarities in the medical and natural resources professions, especially about how internal and external ecosystems respond to their environments. I recommend several references, including my *Watershed Hydrology* and *Conservation of Water and Related Land Resources* books; Guy Murchie's 1961 *Music of the Spheres,* and Davis and Day's 1962 *Water, the Mirror of Science.* A *Handbook of Physics and Chemistry* helped tremendously. I remain inspired, too, by my father who every week talked live for a long fifteen minutes on New York City's WQXR. I sincerely thank all these and you, the listeners. Thanks indeed.

Glossary (and Occasional Symbols)

Acidity Logarithm of the hydrogen ion concentration, a measure of the acidity of water expressed as **pH,** the scale of acidity (1) to base (14), neutral being 7.0. a pH of 8.0 is ten times as basic as 7.0

Acre foot An acre covered with water to the depth of one foot, or 43,560 cubic feet. A flow of one cubic foot per second is equal to about two (2) acre feet per day.

Best Management Practice An engineering and/or managerial means of preventing or ameliorating the pollution effects of land uses—nonpoint sources of pollution.

Evapotranspiration The combination of evaporation (from water surfaces) and transpiration (from vegetation) via stomata, tiny openings (normally) on the underside of leaves, usually abbreviated as E_t.

Interception The interruption of the downward movement of precipitation, usually by vegetation, but also by structures, that results in a redistribution and/or loss of water to the soil and, therefore, to runoff.

Streamflow Discharge or runoff, in some unit of volume per unit of time, for example, cubic feet per second, cubic meters per second, or acre feet per day, often represented by **Q**.

Total Mean Daily Load	The total mass—in pounds and tons, or kilograms and metric tonnes—of a pollutant or nutrient, formerly a point source, but now applied to various land uses (nonpoint sources). Abbreviated TMDL.
Vapor Pressure	The portion of the total atmospheric pressure that is due to water vapor.

Water Trivia

1. It takes 100,000 gallons (379,000 L) of water to manufacture one automobile.
2. 122 gallons (462 L) of water are needed to produce one loaf of bread.
3. It takes 50 glasses of water to grow one glass of orange juice.
4. 97 percent of all earth's water is salty; only 3 percent is fresh water.
5. 3/4 of the earth's surface is covered with water.
6. A 20-minute shower uses 16–20 gallons (60–75 L) of water. A 10-minute shower uses 8–10 gallons (30–38 L) of water. A 5-minute shower uses 4–5 gallons (15–19 L) of water.
7. It takes 3 gallons (11 L) of water to flush a toilet.
8. It takes 30–40 gallons (115–150 L) of water for a tub bath.
9. 10 gallons (38 L) of water is required to hand wash dishes.
10. It takes 20–30 gallons (75–115 L) of water to run a washing machine.
11. The average American home uses an average of 293 gallons (1,110 L) of water a day.
12. It takes 2,500 gallons (9,500 L) of water to produce one pound (2.2 kg) of beef.

Notes, References, and Additional Reading

You can estimate the streamflow in a river pretty easily, although it won't be as precise as if you had a stream gauge or instruments. You need a fairly long straight stretch of stream, for example, fifty or one hundred feet; a stable streambed in that reach; a watch with a second hand, and a measurement or estimate of the depth and width of the stream. First, get the width and average depth of the stream, for example 10 feet wide and 3 feet deep; the cross-section area is 3 times 10 or thirty square feet. Next, at the upstream end of the 100-foot reach, drop some leaves in the water and start the timer. If it takes 40 seconds for the leaves to go the 100 feet, then the *rate* of streamflow (its *velocity*) is 100 feet divided by 40 seconds or 2.5 feet per second. So, the streamflow is 2.5 times 30 or 75 cubic feet per second. Obviously, the figures could be in meters, but the process is the same. If we want to compare the runoff in two streams, we have to divide the streamflow by the area of the watershed, and talk about streamflow in cubic feet per second per square mile (or cubic meters per second per square kilometer).

Calculating Water Rates SB

Block	*Rate*	*Used*	*Total*
1	$ 1.09	4	$4.39
2	$ 3.29	4	13.08
3	$ 9.81	6	58.86
4	$29.43	4	117.72
Grand Total			$194.02

Santa Barbara: 1990
Each block is 100 cubic feet.

Since cooling the air can happen in three different ways, storms come in three types: *cyclonic*—or air mass—storms; *convectional*, the familiar thunderstorms; and storms developing over mountains and or near lakes called *orographic* (from the Greek word meaning "mountain") storms.

There are three forces that act on the atmosphere (and everything else on Earth) to move air around: (1) the temperature gradient force, (2) the pressure gradient force, and (3) the Coriolis force. They will not change with global warming.

Classical, Historical, Not High-Tech, and Readable Resources and Citations*

Ackerman, W. C. 1969. Hydrology becomes water science. *Transactions of the American Geophysical Union* 50, no. 3: 76.

*Bates, M. 1960. *The forest and the sea.* New York: Mentor Books.

*Black, P. E. 1966 (and 2004). *Tall timber and high water.* Adobe Acrobat file available without cost at http://www.watershedhydrology.com. Copyright by PEB also assigned to Archivist at Redwoods National and State Parks, Orick, CA.

———. 1996. *Watershed hydrology*. 2nd Ed. Chelsea, MI: Ann Arbor Press. 449 pp. (Contains more than 800 references to articles, books, journals, and government publications.)

———, and B. L. Fisher. 2001. *Conservation of water and related land resources*. 3rd Ed. CRC Press LLC, 2000 N. W. Corporate Boulevard, Boca Raton, FL. 493 pp. (Contains circa 700 references to articles, books, and manuals on law, organizations, government, policy, economics, and conservation, plus relevant legistation and precedent-setting court cases on water issues.)

Brooks, K. N., P. F. Ffolliott, H. M. Gregersen, and J. L. Thames. 1991. *Hydrology and the management of watersheds*. Ames, IA: Iowa State University Press.

*Davis, K. S., and J. A. Day. 1961. *Water, the mirror of science.* New York: Doubleday Anchor Books. 195 pp.

*Department of Agriculture. 1938. *Soils and men: The 1938 yearbook of agriculture.* Washington, DC: Government Printing Office.

———. 1955. *Water: The 1955 yearbook of agriculture.* Washington, DC: Government Printing Office.

*Forest Service. 1946. *Water and our forests*, Misc. Pub. No. 600, USDA. Washington, DC: Government Printg Office.

Freeze, R. A., and J. A. Cherry. 1979. *Groundwater.* Englewood Cliffs, NJ: Prentice-Hall.

Gaskin, J. W., J. E. Douglass, and W. T. Swank, compilers. 1983. *Annotated bibliography of publications on watershed management and ecological studies at Coweeta Hydrologic Laboratory*, 1934, 1984. General Technical Report SE-30, Southeastern Forest Experiment Station, Forest Service, USDA. Asheville, NC.

Geiger, R. 1957. *The climate near the ground.* Cambridge: Harvard University Press.

Gleick, P. H. 1989. Climate change, hydrology, and water resources. *Rev of Geophysics* 27, no. 3: 329.

*———. 2007. *The world's water 2006–2007: The biennial report on freshwater resources.* Pacific Institute for Studies in Development, Environment, and Security, Island Press, 1718 Connecticut Avenue, N.W. Suite, 300Washington, D.C. 20009-1148.

Hem, J. D. 1970. *Study and interpretation of the chemical characteristics of natural water.* 2nd Ed. Geological Survey Water-Supply Paper No. 1473. Washington, DC: Government Printing Office.

Horton, R. E. 1933. The role of infiltration in the hydrologic cycle. *Transactions of the American Geophysical Union* 14: 446.

*Jenny, H. 1941. *Factors of soil formation.* New York: McGraw-Hill.

Karl, T. R., and K. E. Trenberth. 2003. Modern climate change. *Science* 302: 1719 (and see other associated articles in the "The State of the Planet" series).

*Marsh, G. P. 1874. *The earth as modified by human action.* New York: Scribners.

Meinzer, O. E. 1942. *Hydrology*. New York: Dover.

Mitsch, W. J., and J. G. Gosselink. 1986. *Wetlands*. New York: Van Nostrand Reinhold.

*Murchie, G. 1967. *Music of the spheres: The material universe from atom to quasar, simply explained*, two vols. New York: Dover. 644 pp.

Nace, R. L. 1974. Pierre Perrault: The man and his contribution to modern hydrology. *Water Resources Bulletin* 10, no. 4: 633.

Odum, E. P. 1969. Air-land-water—An ecological whole. *Journal of Soil and Water Conservation* 24, no. 1: 4.

Stumm, W., and J. J. Morgan. 1970. *Aquatic chemistry*. New York: Wiley-Interscience.

Todd, D. K. 1959. *Ground water hydrology*. New York: John Wiley and Sons.

Watson, L. 1998. *The dreams of dragons, an exploration and celebration of the mysteries of nature*. Rochester, VT: Destiny Books. 176 pp. (NOTE: Had I seen this book prior to creating *Water Drops*, I would have either acknowledged Watson's earlier use of the phrase as well as the chapter "The Wonder of Water," or used a different title, or sought permission to use "The Wonder of Water," and given due credit. I did not become aware of this book until after I had completed this manuscript for publication, thus cite the volume here. PEB.)

And on the Internet:

http://www.waterproblems.net/.

http://ga.water.usgs.gov/edu/sos1.html.

http://www.powellmuseum.org/MajorPowell.html.

http://www.pacinst.org/about_us/staff_board/gleick/.

http://hometips.com/home_improvement/water.html.

http://adsabs.harvard.edu/abs/2004AGUSM.U15A..01P.

http://www.islandnet.com/~see/weather/journal/2002/steamfog.htm.

http://www.epa.gov/region7/education_resources/teachers/activities/wateractivity2.htm.

Index